"This book is absolutely urgent today. Everywhere that creation is in crisis. *Following Jesus in a Warming World* offers a warning, a wake-up call, and a way forward for Christians striving to be faithful disciples in the twenty-first century. Rather than pile on the guilt and shame, Kyle Meyaard-Schaap winsomely invites readers to embrace a faithful life that accounts for the ecological state of our world. This book is a balm for the climate-anxious soul!"

Jonathan Merritt, contributing writer for the *Atlantic* and author of *Learning to Speak God from Scratch*

"In a world suffering from ever-higher temperatures, loving one's neighbor has to mean coming to grips with the climate crisis. This faith journey provides both illumination and inspiration for action!"

Bill McKibben, environmentalist and author of *The Flag, the Cross, and the Station Wagon*

"This is a marvelously engaging book about overwhelmingly urgent matters. Kyle Meyaard-Schaap is convinced that we are formed by stories in how we understand reality, and he urges us to approach climate change action in the light of the gospel's Big Story. I pray that many will be moved to climate advocacy by the compelling personal stories that Meyaard-Schaap tells in making his case."

Richard J. Mouw, senior research fellow at Calvin University's Henry Institute for the Study of Religion and Politics

"This book does far more than share compelling facts about a warming world. It tells the story of how following Jesus leads us to protect a world threatened with ecological catastrophe. Kyle Meyaard-Schaap shares the stories of real people on this journey and offers pathways for us to follow. We often say, 'Let's hear the voices of young people shaping our future.' Kyle has empowered them. But those voices are shouting, 'Our future is on fire!' Reading this book doesn't just tell us what to think but shows us what to do."

Wesley Granberg-Michaelson, general secretary emeritus of the Reformed Church in America

"While fellow Christians remain apathetic or dismissive, Christians concerned about the climate crisis can feel they are walking a lonely journey. For these lonely journeyers, Kyle Meyaard-Schaap is a patient, trustworthy, experienced encourager. His irresistible passion calls us back from deceptive narratives into the real story of God's redemptive love for all creation. This book is a deeply scriptural call to advocacy for people and planet as both moral necessity and spiritual discipline. What a gift! Finally, Christians can take courage and hand this book to others, saying, 'This. Read *this*.'"

Debra Rienstra, professor of English at Calvin College and author of *Refugia Faith: Seeking Hidden Shelters, Ordinary Wonders, and the Healing of the Earth*

"Kyle Meyaard-Schaap has helped so many understand that caring for creation is a response to the call of the Creator. In this book, Kyle helps us see anew both the crucial responsibility Christians have in caring for the environment and how issues of environmental promise and degradation can bring us closer to Jesus."

Michael Wear, author of *Reclaiming Hope*

"Kyle Meyaard-Schaap is a storyteller, and in issuing 'a Christian call to climate action' he invites readers to listen to others' stories—to theological stories, to political and historical stories, and to deeply personal stories. Most importantly, by educating, equipping, and inspiring, he invites readers to inhabit a redemptive story. *Following Jesus in a Warming World* is an essential guide for individuals and churches seeking wisdom and loving kindness, and pursuing justice in an era of climate change."

Kristin Kobes Du Mez, author of *Jesus and John Wayne: How White Evangelicals Corrupted a Faith and Fractured a Nation*

"Kyle Meyaard-Schaap is a leading voice calling American Christians to more faithful climate action. This book shares his captivating and convicting journey as an ordinary Christian seeking to love God and neighbor in these extraordinary days of global climate and biodiversity crisis. Read this to be inspired to take a stand and be empowered to find your voice as a follower of Jesus in God's good but groaning world."

Ben Lowe, author of *Green Revolution* and deputy executive director of A Rocha International

"This is an excellent book: insightful readings of the Bible, a perceptive analysis of our culture, and honest and engaging reflections in well-written prose. *Following Jesus in a Warming World* is a must-read for any Christian seeking to understand what it means to live one's faith in these tumultuous times."

Steve Bouma-Prediger, Leonard and Marjorie Maas Professor of Reformed Theology at Hope College

"This is an inspiring book! Kyle Meyaard-Schaap is unfailingly honest, charitable, and compelling, whether he is interpreting Scripture or unfolding the history of evangelicalism or sharing his own journey toward taking seriously the whole story of the Bible and what it means to follow Jesus in this age of climate crisis. Rather than vague platitudes, his advice bears the marks of an experienced leader and activist. Yet what will stick with readers the most is a host of characters whose stories help us see the reality of climate crisis, the power of the gospel, and the beauty of what might be possible when Christians act in joyful and defiant hope."

Jonathan A. Moo, professor of New Testament and environmental studies at Whitworth University

FOLLOWING JESUS IN A WARMING WORLD

A CHRISTIAN CALL TO CLIMATE ACTION

KYLE MEYAARD-SCHAAP

An imprint of InterVarsity Press
Downers Grove, Illinois

InterVarsity Press
P.O. Box 1400 | Downers Grove, IL 60515-1426
ivpress.com | email@ivpress.com

InterVarsity Press® is the publishing division of InterVarsity Christian Fellowship/USA®. For more information, visit intervarsity.org.

All Scripture quotations, unless otherwise indicated, are taken from The Holy Bible, New International Version®, NIV®. Copyright © 1973, 1978, 1984, 2011 by Biblica, Inc.™ Used by permission of Zondervan. All rights reserved worldwide. www.zondervan.com. The "NIV" and "New International Version" are trademarks registered in the United States Patent and Trademark Office by Biblica, Inc.™

While any stories in this book are true, some names and identifying information may have been changed to protect the privacy of individuals.

The publisher cannot verify the accuracy or functionality of website URLs used in this book beyond the date of publication.

Cover design and image composite: David Fassett
Interior design: Jeanna Wiggins

ISBN 978-1-5140-0445-6 (print) | ISBN 978-1-5140-0446-3 (digital)

Printed in the United States of America ⊗

Library of Congress Cataloging-in-Publication Data
A catalog record for this book is available from the Library of Congress.

29   28   27   26   25   24   23 | 8   7   6   5   4   3   2   1

FOR SIMON, AMOS,
AND ALL THE DEAR ONES YET TO COME

Every last bit of it is for you

CONTENTS

INTRODUCTION

"I'M A VEGETARIAN."

His words clanged across the chasm between us. Vague images of hemp friendship bracelets, vegan pizzas, and red paint dripping off fur coats swam across my mind. I couldn't name one person from my conservative Christian school or church who was a vegetarian.

He had always been a hero of mine, my ever-present playmate through childhood, my guide through the mysteries of adolescence, my friend and my role model. Now my older brother, newly returned from a semester abroad, was telling me he was someone totally different—someone I didn't recognize.

Along with our older sister, my brother and I had grown up together in a tight-knit Christian community. The Christian K–12 school we attended was woven into the fabric of our town, and church steeples rose above every neighborhood. We were invested in our local church, attending Sunday services as well as several other programs throughout the week. We had been told the stories of Jesus before we could walk. We knew the names Abraham and Moses before Cinderella or Snow White. By the time I was in high school, I was on my church's praise and drama teams, I was a student leader in our youth group, and I served as a summer student intern implementing neighborhood outreach programs for kids in the surrounding community.

All of this had formed my siblings and me in profound and particular ways. We learned about the overwhelming love of God, the free gift of grace, and the beauty of Christian community. Our worldviews were shaped by biblical values like faith, hope, and love. Our measuring stick for integrity involved patience, kindness, self-control, and the rest of the fruit of the Spirit (Galatians 5:22-23). And yet, such a seemingly innocuous announcement from my brother about a simple lifestyle choice was enough to send me reeling—to make me feel unmoored, lost, and betrayed. Why?

My brother had not attended a clandestine program designed to deconstruct his Christian worldview and erect in its place a monument to secular humanism. Though foot soldiers of the culture wars like to imagine such Trojan horses lurking around every corner of higher education, the study-abroad program he had attended was thoroughly Christian, complete with regular worship and courses designed to bring Scripture and theology into conversation with ecology and environmental studies.

When I had the ears to hear him, the descriptions of his learning and growth mapped perfectly onto the Christian values and worldview we had already been given. For him, becoming a vegetarian, joining the Sierra Club's mailing list, and memorizing the bus routes to and from his Christian college campus did not constitute a rejection of our Christian values but were an expression of them. He had not jettisoned his Christian identity. He was living more fully into it.

So why was I so unprepared to accept these choices? Why was I so blind to the deeply Christian values informing them? Why did I have memory verses at the ready to spark thankfulness, to offer comfort, and to cultivate trust but none that called me into deeper relationship with God's creation or helped me make sense of the existential crisis of climate change? What had my Christian formation omitted, and why? Which scriptural passages and themes had been sidelined? What biblical truths had been suppressed? Did my commitment to following Jesus actually permit me to care about things like species extinction, pollution, and climate change? More than that, did it require it?

After my brother first forced me to confront these questions almost fifteen years ago, they have consumed me. They have taken me across the world, from the maize fields of Kenya to the bayous of Louisiana, from the mountaintops of West Virginia to the vaunted halls of international climate negotiations in Paris. They have forced me onto my knees in prayer and propelled me into the streets in protest.

It is these questions that this book seeks to answer. Across my hundreds of conversations with young Christians about climate change over the last decade, one theme looms above all the rest: silence. Countless millennial and Generation Z Christians—particularly those who were formed by White, conservative evangelicalism[1]—simply haven't been told that their faith has anything to say about such questions. Those of us who have come to Christian climate action have come to it circuitously, via some combination of lonely epiphany, painful deconstruction, and growing isolation from friends and family who view our Spirit-breathed conviction as political radicalization rather than Christian discipleship.

This book is for every Christian who has walked this lonely journey. It is for every Christian who has looked out at a world ravaged by the impacts of climate change, at the inability of their seemingly oblivious faith community to do or say anything about it, and quietly wondered if they're losing their mind.

The need for the church to wake up and act is urgent. Since 2004, seven major studies have been conducted to measure the level of scientific consensus among climate experts. Depending on the various methodologies used, these studies have found that anywhere from 90 to 100 percent of climate scientists agree that the climate is warming at an alarming rate—about a hundred times faster than natural warming trends in the past—and that the primary cause is the extraction and burning of fossil fuels on an industrial scale. The average level of consensus across all these studies is 97 percent. What's more, the studies found that the higher the level of climate expertise among those surveyed, the higher the level of consensus on human-caused warming.[2]

This makes sense. While the odd geologist or chemist may have their doubts about human-caused climate change, they don't have the same level of training and expertise in the field as physicists and climatologists. A podiatrist *might* be able to diagnose a cancerous tumor, but wouldn't you rather trust an oncologist?

Yet, research shows that even though 97 percent of climate scientists agree that climate change is real, it's bad, and it's us,[3] the US population remains largely in the dark about the scientific consensus on climate change. Only 59 percent of US adults understand that scientists "agree" that global warming is happening, and only 24 percent know that the scientific consensus on climate change is over 90 percent.[4] This matters because on complex issues that require significant expertise—like medicine, law, or the climate system—people defer to experts. Studies have shown that when people understand that experts are unified in their findings regarding climate change, people are more willing to support policy solutions to address it.[5]

This consensus has been a long time coming because, in science, consensus is achieved slowly. It emerges over time as thousands of studies are conducted, peer reviewed, replicated, and either rejected or affirmed. Very slowly, common results begin to rise to the surface as certain results are tested and affirmed and others are tested and found lacking.

Many skeptics of the scientific consensus of climate change accuse climate scientists today of "group think" and want to see more robust scientific debate and disagreement in pursuit of the truth. These arguments ignore the simple fact that the scientific debate they crave has already occurred. The study of the composition of Earth's atmosphere, the properties of heat-trapping gases, and the warming of Earth's surface dates back to the middle of the nineteenth century. Scientists have been having their debate for over 150 years, and the results are in.

This uniform alarm among the world's leading experts, achieved slowly over a century and a half of rigorous study, is reflected in their increasingly dire warnings to the world. The Intergovernmental Panel on Climate Change (IPCC)—a global, independent body of scientists from all over the world—has been issuing reports since 1990 on the state of climate change and what policymakers must do to avert its worst impacts. As climate science has gotten more precise and policymakers have continued to do little to slow climate change's death march, the warnings have gotten more strident.

In 2015, when the landmark Paris Agreement was reached at the UN Climate Change Conference (COP21), the world agreed to limit "the increase in the global average temperature to well below 2°C above pre-industrial levels."[6] This represented a compromise between more ambitious nations and those who would have preferred to keep the target vaguer. However, due to the creative activism of nongovernmental observers to COP21 and the tireless advocacy from small island nations, the agreement also committed to pursue efforts "to limit the temperature increase to 1.5°C above pre-industrial levels, recognizing that this would significantly reduce the risks and impacts of climate change."[7]

This more ambitious target was affirmed in 2018 when the IPCC released an updated report, just three years after the herculean diplomatic effort of striking the Paris Agreement. The new report said that a 2° Celsius rise was still unacceptably dangerous and that government and industry must do all that they can to achieve the lower target of 1.5° Celsius. This would require "rapid and far-reaching" transitions in energy, land, infrastructure, and industry on a scale never before seen in human history.[8] Yet, according to the World Resources Institute, current emission-reduction commitments from the world's nations aren't up to the task and put the world on a trajectory for a 2.4° Celsius rise by the end of the century.[9] This would be bad enough if countries were on track to meet these commitments. Instead, current policies in place are projected to lead to a global temperature rise of 2.7° Celsius by 2100.[10]

What's more, the report found that we have less time than we originally thought. The Paris Agreement laid out a plan for countries to peak their greenhouse gas emissions by mid-century, and then spend

2050–2100 reaching net-zero emissions—striking a balance between emissions that throw greenhouse gases into the atmosphere and sinks (i.e., forests, oceans, soil) that draw them back down and lock them away. The report found that the world needs to begin reducing its emissions far more rapidly—halving emissions by 2030 and reaching net zero not by 2100 but by 2050. If we do not, warns the IPCC report, we will already begin to see catastrophic impacts and irreversible tipping points by 2030.

How we respond to this reality as followers of Jesus is of enormous spiritual, personal, and existential importance. It is nothing less than the defining moral challenge of our time. Through theological and scriptural exploration, storytelling, and practical steps to turn our faith into meaningful action for climate justice, this book will show that acting to avert and adapt to climate change is a fundamental part of following Jesus in a warming world, and that the church is uniquely equipped to respond to the climate crisis with hope and action.

Finally, a brief housekeeping item. Throughout what follows, I use the traditional masculine pronouns for God. This is purely for readability. Scripture explicitly ascribes both masculine and feminine attributes to God. It calls God Father, and it also compares God to a mother hen gathering her chicks (Luke 13:34). All human language used to describe God is analogical, which means we should always hold all God-talk loosely. This is especially true of socially constructed categories like gender. There is nothing essential in God's nature that requires God to be male—the Godhead transcends gender. Yet, for better or worse, our human language about God is gendered. I therefore observe the traditional practice of using masculine pronouns for God, not because masculinity is essential to God's nature but because it makes for much easier reading (and writing).

It took me a long time to recognize my brother again, and to cut through the fog of unconscious assumptions and biases that I had inherited. These were biases that had me believe that authentic Christian faith and a concern for the well-being of all God's creation were dissonant, if not mutually exclusive, and assumptions that led me to believe a commitment to following Jesus called me only to a narrow set of political concerns, and that climate change lay firmly outside of it.

Thank God that the gospel and its good news for the world are so much bigger than the lines we draw around it.

1

COAL AND THE GREATEST COMMANDMENT

WE MET AT A GAS STATION off the county highway at sunrise. There was no mistaking which pickup truck was his. It was plastered with bumper stickers that carried phrases like "Friends of the Mountains," "I Love Mountains," and "Topless Mountains are Obscene." If any doubt remained, it was quickly dispelled when he hopped out of the driver's seat to greet our group. All of five feet tall, with his signature faded overalls and straw brim hat, Larry Gibson was one of a kind.

It helped that our group knew who to look for. Larry's reputation preceded him. Inspired by the likes of Larry and others, our small group of a dozen or so college students from Grand Rapids, Michigan, had decided to trek to the hollers of West Virginia and to spend our spring break not on some sandy Gulf Coast beach but in the coal fields of Appalachia. We were there to serve in whatever imperfect and halting ways we could, but mostly we were there to learn, to bear witness, and to lament. We had come to rural West Virginia to experience the practice of mountaintop-removal coal mining.

Mountaintop-removal coal mining is a particularly harmful form of strip mining that consists of clear-cutting large swaths of old-growth Appalachian forests from the top of mountain ridges. In much of Appalachia these are some of the oldest forests in the world. Once the trees are cut and cleared, holes are bored at intervals into the mountain crust and dynamite is planted deep in the heart of the mountain. The dynamite is then exploded, erasing up to six hundred feet of ancient elevation in an instant. The rubble that's left behind, called "overburden" in industry jargon, is then cleared away.

Thanks to a 2002 change to the Clean Water Act, this waste is officially classified in the same category as soil and dirt, allowing the mining companies to simply push the mix of rubble, ash, and toxic heavy metals off the side of the mountain into the streams and rivers below. Since this rule change, this "valley fill" has buried more than two thousand miles of headwater streams and polluted many more.[1] All of this in order to access thin seams of coal so near the mountain's surface that traditional deep-well mining can't get at it.

These coal seams are marginal. Before technology advanced to the point where they could be harvested, they would simply be left alone. Because of this, the profit margins for mountaintop-removal operators are tight. That's why mountaintop-removal sites employ one worker for every eleven workers employed at traditional deep-well mining sites. Like the perennial push to develop the Arctic National Wildlife Refuge— established by a Republican president in 1960, expanded by a Democratic president in 1980, and meant to serve as a guardrail for our collective craving for cheap oil and gas—the practice of mountaintop removal has the feeling of a drug user working harder and harder to get a fix. Our collective addiction has driven us to ever more extreme behavior. As with other addictions, the collateral damage of our singular drive to score is profound.

We had come to see Larry because Larry understood the logical conclusion of this kind of runaway addiction, and he had committed his life to putting a stop to it. By the time we met him in Kayford, West Virginia, on a chilly, fog-filled day in the spring of 2012, the consequences of our addiction to cheap coal were everywhere. Mountaintop removal had scarred thousands of square miles of Appalachian landscape, displaced neighbors who had called now-buried Appalachian hollers home for generations, and poisoned countless neighbors living downstream. We had read about it. We had watched documentaries about it. We had discussed it as a group ad nauseam. Now Larry was going to help us see it with our own eyes.

Gravel crunched under Larry's tires and fog swirled in the headlight beams as we inched our way up the steep access road to Larry's cabin. I tried not to think about the sheer cliff face inches from the passenger side tires, made all the more menacing by its camouflage of thick fog. I looked over at Larry, who was completely at ease traversing the treacherous mountainside. To take my mind off my proximity to certain death should Larry's hand slip from his steering wheel, I listened to the police scanner mounted on the inside of Larry's windshield. Chatter filled the cabin in blips and chirps. My ear, unaccustomed to the distorted words carried to us on the wind, had a hard time following much of it. One word, though, was uttered so frequently that I began to pick it up.

"Who's Stickers?" I asked Larry.

"Me," he replied. "That's their nickname for my truck. They're tracking us."

Larry explained that the voices crackling through the scanner belonged to a handful of men who had been hired by the coal company in town. Like many of the mountains surrounding Larry, shallow veins of coal crisscrossed the mountain on which Larry's home sat. When these were discovered, the local coal company went about purchasing the mineral rights from the inhabitants of the mountain in order to blow it apart—to tap the coal seams and bleed the mountain of its precious payload. One by one, Larry's neighbors sold their rights and moved off the mountain.

Not all the residents of Kayford Mountain wanted to sell. When the coal company failed to entice them with larger and larger carrots, they resorted to using sticks to drive them off their land. Intimidation campaigns harassed neighbors until, frustrated and frazzled, they sold their rights and escaped the abuse. By the time we visited Larry, he was the last holdout. The full force of the intimidation campaign was now trained exclusively on him.

"Maybe he'll drive off the edge this time," chirped a voice from the scanner.

"Fingers crossed," crackled the reply.

Larry's cabin finally loomed out of the fog as we reached the mountain summit. It bore signs reading "We are the Defenders of the Mountains" and other paraphernalia from his decades-long fight on behalf of his and others' homes. Welcoming us inside, he showed us the bullet holes left by his intimidators. "Sometimes they get drunk and like to come have a laugh," he explained. As if to prove beyond doubt what had caused the holes, he showed us the bullet casings he had collected over the years.

After warming ourselves by his wood-fired stove, we set out on foot for the main event, the reason we had come to see Larry: a mountain ridge not far from Larry's cabin that offered a panoramic view of a mountaintop-removal mining operation on the mountain next door. As we walked the steadily inclining path to the ridge, our feet kicked and stumbled over coal seams so close to the surface that they looked like black tree roots. Larry shared stories of Kayford Mountain with us as we walked: neighbors long since moved away, property long since sold off. He told us about Kayford Cemetery—home to generations of revered patriarchs, quirky uncles, beloved mothers and grandmothers—now buried under the rubble of a neighboring mountaintop-removal operation and unreachable by those relatives who still lived and who ached to visit those sleeping beneath the dust and debris. He spoke of Kayford Mountain as a beloved.

Suddenly, Larry threw out his arm like a parent stopping short at a red light. We had reached the ridge. The fog that had surrounded us on our summit up the mountain still hung like a veil across the expanse between us and what we had come to see. Pictures had primed us for what to expect: inert, neatly terraced earth rather than sloping, vibrant mountainside rising to the sky. We were prepared to encounter a pallet of drab grays and browns where once there was wild, vivid green; a broken and scarred moonscape sprawling across hundreds of acres of decimated ecosystem.

I saw none of it. I saw only a thick blanket of fog pressing in on all sides. Larry, perhaps sensing our disappointment, instructed us, "Forget your eyes. Use your ears." I listened hard. I heard absolutely nothing, no cacophony of birds calling to one another from the branches of Appalachia's old-growth pines, no abundance of life that should have been pulsing to us like an electric current across the expanse. Instead, I heard the vast emptiness left by the thousands of feet of ancient elevation now leveled forever. I heard nothing out of the void. Nothing at all. I never knew silence could be so deafening.

Our visit with Larry comprised just one of our seven days in Kermit, West Virginia, and the nuns who hosted us kept us busy. We accompanied Sister Kathy on her weekly trip down the mountain to the local school where she volunteered and heard one teacher tell us, "If you ask any of these kids what they want to do when they grow up, each one of them will tell you the same thing: mine coal. It's all they know."

We visited the neighbors on the mountain where the nuns lived, helping to split wood and sipping iced tea on front porches. We listened as wives told us about their husbands' inoperable tumors and how much time they had left. A mother recalled the latest trip into Charleston to treat her eleven-year-old daughter's ovarian cancer—a reality for countless people downwind and downriver from mining operations in coal country. The nuns would comfort and pray for their neighbors while we looked on, intruders on the community grief.

We were hosted for a day by the coal company that mined the region surrounding the nun's mountain (not the same company terrorizing Larry). They welcomed us into their spacious conference room, and while we ate a catered breakfast, they shared with pride and conviction how they and the people of the area had partnered for decades to keep America's lights on. We donned hard hats and reflector vests and descended into a deep-well mine. We piled into pickup trucks to see a "reclamation site"—a retired mountaintop-removal mining site with spotty patches of scrubby grass, which the coal company reps tried to convince us was just as good as it was before they had blown it all to smithereens. After dropping us off at our cars at the end of the day, one of the reps said, "I hope this helped give you some appreciation for the good that coal has done." Driving away, I guessed that our suspicion of them had not been well hidden.

We helped at a food drive at a small country church nestled in a nearby holler. The growth of machine-driven mountaintop-removal mining in the region and the inexorable decline of the coal industry meant there was no shortage of need. Out of work and underemployed coal miners lined up to receive their food with defiance, shame, and resilience mingled on their faces. Volunteers handed out boxes of food alongside us. When the rush had diminished, we sat down with a few of the volunteers and the pastor of the church, all of them previous coal miners themselves. We shared about our experience and what we had learned over the course of the week. We talked about cancer and bullet holes, police scanners and economic despair. The people of the holler, explained the pastor, all had ties to coal. If they didn't mine it, their daddy or their granddaddy had. Coal mining was their heritage, their livelihood, their identity. We left with the distinct impression that though the pastor and volunteers sympathized with Larry's plight, his was a story they had heard all too often. To them, Larry's predicament was simply the unfortunate yet necessary cost of business. "After all," the pastor said to us toward the end of our time together, "God gave us the coal to bless us. He wants us to use it!" His tone made it clear that, at least for him, this settled the matter completely.

Coal and the Greatest Commandment

In the Gospels, Jesus is asked, "What is the greatest commandment?" Depending on the Gospel account, the questioner is a Pharisee (Matthew 22:36), a scribe (Mark 12:28), or a lawyer (Luke 10:25). In other words, a religious insider with intimate knowledge of the commandments himself. According to Matthew and Luke, the question is intended as a test. It's easy to see why. In Jesus' day, there were more than six hundred recognized commandments found in the Hebrew

Scriptures, and there were various Jewish sects that interpreted these commandments differently.

The Sadducees tended to interpret them broadly and attempted to contextualize them to a world where they found themselves under the thumb of capricious Roman occupying forces. The Zealots applied the Law in the context of their occupation in a way opposite to the Sadducees. Whereas the Sadducees cozied up to the Romans in order to secure for themselves some modicum of control and autonomy, the Zealots sought to violently overthrow Roman forces through guerrilla warfare, assassination, and violent uprisings. The Essenes, privileging the commands to be a holy and set apart people, separated themselves entirely from society, founding communities in the wilderness, where they pursued purity and holiness together.

The Pharisees were, by most accounts, the most legalistic of the sects and had even created what's been called a "fence around the Law," a complex system of countless additional rules and regulations as insulation for the original, God-given commands. The theory was that even if they broke one of these extracanonical rules (e.g., exceeding a maximum number of allowable steps on the Sabbath), they would still be able to keep the actual commandment found in the Torah ("Everyone is to stay where they are on the seventh day; no one is to go out," Exodus 16:29). This system of auxiliary rules gave rise to complicated taxonomies and split various schools of thought about how best to organize, rank, and privilege them. Rabbis offered differing opinions, splitting the Pharisees into opposing philosophical camps.

The question put to Jesus in the Gospel account is dripping with cultural, social, and political baggage. Jesus' questioner is essentially forcing him to take a side, guaranteeing that he will alienate at least one faction of his listeners. "Which school of thought do you subscribe to, Jesus? Which tribe is yours? Team Rabbi Hillel or Team Rabbi Gamaliel?" It isn't hard to recognize the same kind of binary thinking at play in our own context today. Wherever we turn, we hear similar questions put to us when it comes to climate change: Which camp are you in? Skeptic or believer? Liberal or conservative? Blue or red?

Tragically, this kind of dualistic thinking seems just as prevalent in the church as anywhere else in society. An ingrained sense of moral certitude in the midst of a complicated world seems to be the inheritance for those of us who grew up in late-twentieth-century evangelical culture indelibly shaped by the religious right. Questions of right and wrong, biblical and "worldly," were easily discernible to those appropriately submitted to the gospel. In a political, religious, and social world shaped by certainty, moral nuance is often flattened out into clear-cut answers of right and wrong. Ambiguity is not merely a nuisance but a threat to a worldview that asserts that the Bible and its intentions are straightforward and that moral behavior is self-evident. Complexity is ignored and discernment is discouraged in favor of strict moral directives. Dos and don'ts dominate the believer's moral imagination.

That's why Jesus' answer to this question is such a gift. He refuses to take the bait. He refuses to accept the binary terms of the question and instead offers something that is at once simple and deeply profound: love God and love your neighbor. That's it. There are no complex legal categorizations or hierarchies, no endorsement of one camp over another. Instead, what he provides is a radically faithful distillation of discipleship. The entirety of our response to the story of God's saving work in the world, says Jesus, can be summed up like this: love God with everything you've got, and love your neighbor as if their present circumstances and future prospects were your own.

How refreshing it is for us as followers of Jesus living in a polarized age to see this kind of creative resistance from Jesus, to see him refuse to play by the rules of zero-sum tribalism and instead chart a third way forward rooted in love and grounded in God's Word. When it comes to climate change (and any other pressing social, economic, or political issue), we can feel enormous pressure to choose a side. We feel we need to align with one group over against another. There is pressure to flatten out the complexity and nuance of climate action by mashing it together with a whole host of other policy positions privileged by one political party or the other.

What a gift Jesus offers to us, then. A clarifying lens that transcends petty games of us versus them, that frees our imaginations from the shackles of zero-sum politics, and that reminds us that love is our highest calling and that in God's economy of abundance, there is plenty to go around.

After all, how can we love our neighbors well if we remain silent in the face of circumstances that threaten their livelihoods and poison their bodies? How can we tell our brothers and sisters in Christ, "I believe you," when they describe the ways that climate change is harming them and their families, and then do nothing to try to change their circumstances? How can we love our neighbors without fighting for their right to clean air and water, and a safe and stable world where they can flourish and thrive? How can we be pro-life in a warming world if we ignore the myriad ways in which climate change endangers and extinguishes life?

The Bible has a word for the kind of faith that sees the suffering of its neighbors and does nothing to respond: dead (James 2:17).

⁓

All the people I met and the stories I heard on that trip to West Virginia complicated my perspective. I had gone fully convinced of the horrors of mountaintop-removal mining and of the need to transition away from fossil fuels as quickly as possible. My time with Larry, my up close and visceral experience of mountaintop-removal mining, the stories of pre-teens losing their hair from chemotherapy, and the long lines of abandoned coal workers had confirmed my belief. I had been ready for that.

What I hadn't been ready for was the pride with which people spoke about coal mining and their role in powering America. I hadn't been ready for the sparkle in the eyes of the school kids who spoke animatedly about the day they would be old enough to join their parents in the mines. I hadn't been ready for the desperate economic reality that held the communities we visited in a stranglehold. As one local told me, "If you ain't working in the coal mines, you're working at the Dairy Queen."

In some ways, my strident certitude about the immorality of fossil fuels had been chastened. It was complicated. People's lives were wrapped up in fossil fuel extraction, transportation, and distribution. Human faces now swam across my vision when I considered concepts such as mountaintop removal, environmental justice, and a just transition away from fossil fuels toward cleaner alternatives.

Yet, as empathy began to soften my outlook, my moral clarity was being hardened. Pride in the role that Appalachia has played in driving the Industrial Revolution and bringing millions of people out of poverty over the nineteenth and twentieth centuries can be celebrated. We can honor the contributions that fossil fuels have made to American prosperity and well-being without giving fossil fuels a free pass in perpetuity. We can acknowledge both that coal has unlocked tremendous economic growth and that it is simultaneously endangering that growth by driving dangerous climate change. We can recognize that coal has put food on millions of American tables *and* poisoned the drinking water. Both can be true at the same time.

I often see the faces of Larry, the nuns, those out-of-work coal miners, and so many others I've met throughout the years. I carry them with me. I do this partly because I promised to—I promised to tell their stories to anyone who would listen. Mostly, though, I recall these faces because I can't help it. Larry's wide-brimmed hat, the nuns' generous spirit, those kids' sheepish smiles—I couldn't shake them if I tried.

And I wouldn't want to try. Each one is a gift. Each story is a kind of sacrament, a physical manifestation of an invisible grace. I don't want to forget the tenacity and fierce righteousness of Larry's cause in the face of gross injustice, the strength it takes a mother to tell the traumatic medical history of her young daughter to a group of total strangers, and the resilience of an entire community in the grip of economic despair. All of it holds out to me truths about God's character and about the shape of our collective calling to follow after Jesus in whatever world we may find ourselves—even, and perhaps especially, a warming one.

All of it is a reminder that the partisan rancor, the esoteric legislative jargon, the layers of bureaucracy, and the impenetrable technological minutiae that make up the constellation of actions necessary to stave off climate catastrophe are only part of the picture of Christian climate action. Underneath all of it are people, just people. People who are living, breathing image-bearers of God doing their best to stay alive, to stay healthy, and to keep safe the ones they love.

Larry and the others whose stories I carry in my heart are a sacramental reminder of the holy formula that creation care equals people care. They are a living, breathing reminder that to be pro-life in a warming world requires that we be actively, aggressively anti–climate change. They are reminders that the way we live and move and have our being on the earth directly impacts others' ability to do the same. When we receive these sacramental gifts with the ethical posture of love held out to us by Jesus, suddenly the petty partisan nonsense that so often surrounds discussion about climate change falls away. In its place is a radically simple question: Will we love?

Larry Gibson died of a heart attack in 2012, mere months after our visit. He died with his beloved Kayford Mountain still in desperate danger. I think of him often, and all the other people I met in West Virginia. I think about the teachers at Kermit Area School, trying their best to keep their students fed, clothed, cared for, and, if there was still time, reading at grade level. I think about the neighbors fighting pediatric cancer and Black Lung Disease. I think about the mining executives doing all they could to maximize profit for their shareholders, and the miners themselves who swelled with pride at the mention of powering America. I think about the country pastor and the food pantry volunteers, sympathetic to Larry's plight but resigned to it as a necessary cost of powering the economic engine of America. I have no doubt that love motivated many of them: love for their students, love for their work, love for their church, and love for the neighbors they could touch and see and hand a box of food to.

But I can't help wondering: What if love could have done more for Larry? What if rather than a realpolitik pragmatism or a resigned fatalism, mountaintop-removal coal mining in Larry's community were seen first and foremost through the lens of love? What if it could be seen through the lens of what neighbors owe to each other, of what we owe to the world that sustains us, of what we owe to the Creator who calls it all good?

If the creative, third-way love of Jesus were our ethical lens, I think Larry might have died differently. Rather than isolated and alone in his fight, with bullet holes in his cabin, he might have been assured and at peace that his fight for Kayford would carry on without him. Maybe the theology of the pastor we met would have been shaped less by the prevailing economic and political forces benefiting from the abuse of the land around him and more by a vision of love and protection for God's creation. Maybe he could have recognized the birds, the mountains, the streams, Larry, and all the people negatively affected by the company's practices as objects of the love to which Jesus calls us and could have acted out of that love rather than rationalizing the problem away. Maybe after handing out boxes of food, he and his volunteers would have driven the short distance to Kayford to stand alongside Larry in his fight for dignity for himself and for the mountain that he loved and called home.

What if Christians across the United States were increasingly formed, week after week, into people who saw the created world not as inert raw material meant for nothing more than powering our industrial machines but as the Creator's masterpiece, shot through with the glory of God, with a destiny of its own in God's coming good future? In other words, what if US Christians were formed to love God's world and to love the people who depend on it for their survival? Then maybe the local churches surrounding Kayford Mountain would have protected Larry from the harassment he suffered. Maybe they would have protected the mountains surrounding Kayford that had already been toppled. Maybe Christians across the country would have known Larry's story and would have marched in the streets and knocked down the doors of Congress demanding a stop to the abuse. Maybe even today, years after Larry's death, Christians would be exerting persistent pressure on corporations to find alternative means of energy production that treat humans like Larry and the rest of creation with dignity and respect. Maybe they would be leading the charge to transform all of humanity's relationship to the created world by protecting endangered species, eliminating dangerous pollutants, and stopping climate change in its tracks.

If the American church had been formed by this kind of interpretation of Jesus' command to love God and our neighbors alike, maybe Larry could have died differently. Maybe he could have been surrounded by neighbors long since driven out. He might have had the sound of hymns in his ears sung by members of the local church standing guard at his door. He could have known that his kinship with Kayford Mountain was not an isolated relationship but was shared by Christians the world over in all their varied, wild, and beautiful places.

2

HOW DID WE GET HERE?

WHY HAS THE AMERICAN CHURCH struggled so mightily to understand that protecting and preserving creation is a fundamental part of loving God and neighbor? Why was my brother's decision to cut out meat and start taking the bus an existential crisis for me? Why are so many of the young Christians I've met integrating their faith with climate action outside the walls of their local church and then keeping quiet about it on Sunday mornings? Why is the topic of climate change in the church anathema in so many corners, and simply ignored in so many more?

The causes are myriad, and we'll spend the chapter to come exploring some of them. Yet, at root the primary reason the US church—and especially the evangelical portion of it—has been silent on climate change for so long is because it has allowed itself to be formed by particular stories. Stories that serve to consolidate political power. Stories that overlook major scriptural truths about the central role of creation in God's work of redemption. Stories that have convinced Christians to view scientific discovery not as a divine gift but as a threat. Stories that have been created and peddled in the name of profit rather than truth.

Of course, we are all formed by stories that shape our realities and circumscribe the boundaries of what is possible and what is not. Many are positive stories about the love and sacrifice of ancestors told around holiday tables. Others express the boundless love and grace of God told through mouthfuls of bread and wine or soak us to the bone with the waters of baptism.

There are other stories too. Centuries of racist narratives and the unequal structures designed to justify them have convinced generations that arbitrary differences in the natural variability of human appearance are in fact God-designed signals of immutable inferiority and superiority. Sexist stories across time and space and spanning human cultures have fed the lie that women are inherently weak, feeble-minded, and ill-suited for leadership or positions of power. Other cultural stories have perpetuated similar lies about the inherent unworthiness of people with disabilities, LGBTQ+ people, immigrants, refugees, and other "outsiders."

Whether we know it or not, we also hear stories about climate change. These stories teach us that climate change is too political to risk bringing into the church. We hear that people taking action to address climate change aren't "like us" and that they don't share our values. We are led to believe that they are probably out to pull one over on us—to steal our freedom, our money, or both. Or maybe most pervasive is the story that climate change matters so little to Christians that it's not worth talking about at all. In short, when it comes to climate change, we have been formed by stories other than the story of God.

The Stories That Form Us

Good Christians vote for Republicans. When I was in fifth grade, my school held a mock presidential election. It was the year 2000, and outside the walls of my Christian day school a presidential campaign was in full swing. For months, the country had been saturated by campaign ads and endless news loops about the latest triumph or gaffe from the campaign trail. George W. Bush—the plain-spoken, born-again governor from Texas—and Al Gore, the Tennessee senator turned eight-year vice president, were each vying for their shot at the Oval Office.

As eleven-year-olds, we understood little of the issues or the politics. Yet, just like everyone else in the country, we had been exposed for months to TV ads whose content we barely understood but whose tone was unmistakable. We had soaked up the chatter from talk radio pouring from our parents' car speakers to and from school. We had imbibed the implicit assumptions of our larger communities. All of it had primed us for the day's event.

On the day of the school election, after all the ballots had been cast first thing in the morning and the ballot boxes were walked down to the school office for tallying, our teachers did their best to turn our focus toward the mundane tasks of that day's lesson plan. Yet, it soon became clear to everyone that fractions were going to have a hard time holding our divided attention. Every chance we got that day, we exchanged intel and conducted exit polls. Balls were left lonely on the recess blacktop as groups of students huddled to share their votes with one another. Lunch conversations were dominated by analysis and punditry. As the day wore on, trends and patterns began to emerge. A consensus was building.

As the clock beat its interminable path to 2:45 p.m., even the teachers abandoned their lesson plans and indulged in the excitement of the impending announcement. Students roamed freely among the desks, sharing their final hot takes with anyone who would listen. Finally, minutes before the day's final bell, the intercom system crackled to life and our principal's voice boomed across the school. "Boys and girls, it has been an exciting day. Thanks to all of you for participating in Pine Ridge Christian Elementary's 2000 presidential election. The votes have been tallied, and the students have spoken." Every student leaned in and held their breath. "With 97 percent of the votes, the next president of Pine Ridge Christian Elementary School is George W. Bush!"

The announcement hung in the air. A silence settled over my classroom. Students exchanged looks of shock. This was not the result we had anticipated. Finally, a student in the back of the room uttered the burning question on all our lips. "Who the heck voted for Al Gore?!"

The anticipation that had marked that day was rooted not in genuine curiosity about who would win the election. In contrast to the national election a month later, which would prove to be one of the closest in American history, the winner of the Pine Ridge Christian Elementary School's presidential election was never in doubt. The question being asked in hallways and bathrooms that day was not who would win but rather if anybody could fathom the possibility of Al Gore getting any votes. By the end of the day, a consensus had emerged among the students, forged by group think and echo-chamber analysis. It was inconceivable to almost every student that George Bush would receive any less than 100 percent of the vote. Christians voted for Republicans. Everybody knew that.

The reasons why a bunch of fifth graders in Holland, Michigan—and millions more Christians growing up in the 1980s, 1990s, and early 2000s across the United States—internalized this very specific political orthodoxy are many and complex. Some apologists for the alliance between Christian faith and Republican politics will argue that it is simply the result of a groundswell of grassroots resistance to the moral outrage of abortion and the Republican Party's willingness to end it. Critics of the religious right will point to evidence that Southern fundamentalists and hard-right partisans made common cause in the late 1970s, not out of concern for abortion but out of the racist impulse to allow Christian schools to practice segregation while continuing to enjoy exemption from federal taxes. Both are true, yet neither is the full picture.

Beginning in the 1960s and 1970s, grassroots evangelicals did begin making common cause with Catholics in the fight against abortion. For their part, evangelical and fundamentalist leaders like Billy Sunday, John Rice, and Bob Jones Sr. were already speaking out against abortion before the *Roe v. Wade* decision of 1973. They founded anti-abortion organizations and supported legislation to overturn *Roe* in the first half of the 1970s. While it would take some time for abortion to become the partisan issue that it is today, by the time the Republican Party adopted its first anti-abortion position in 1976, a nascent but growing anti-abortion movement of conservative, evangelical Republicans was beginning to coalesce.[1]

At the same time, many of these same evangelical leaders were fending off attempts by the IRS to revoke the tax-exempt status of their religious schools for continuing to practice segregation in violation of the *Brown v. Board of Education* decision of 1954. They got help from an insurgent movement within the Republican Party called the New Right. Together, they tapped into the growing anti-abortion grassroots energy to galvanize the political movement known as the religious right.[2]

Yet, to tell both of these histories so cleanly, as many like to do, divorces them from the larger social and cultural context of the day. These decades in the United States were a time of profound social change. The civil rights movement, second-wave feminism, the anti–Vietnam War movement, the growing gay rights movement, and more were upending the traditional social and cultural norms that US evangelicals took for granted. Their growing anti-abortion efforts were part of a larger resistance to changing cultural expectations around gender and sexuality and were coupled with organized resistance against the Equal Rights Amendment and local gay and lesbian nondiscrimination ordinances. Similarly, evangelical efforts to protect the tax-exempt status of their schools were part of a larger backlash against the gains of the civil rights movement. In other words, to tell simple histories of the religious right that focus *only* on abortion or *only* on race misses the larger contextual landscape out of which the religious right emerged. It also ignores other key issues that helped forge the alliance between evangelical faith and conservative politics in the final decades of the twentieth century, including—crucially, for our purposes—energy and environmental policies.

According to historian Darren Dochuk, oil is foundational to the story of America. "Much more than a material form," says Dochuk in his book *Anointed with Oil*, "oil is an imprint on the American soul."[3] And from the beginning, the pursuit of oil in America has been inseparable from Christianity.

By the 1890s, John D. Rockefeller's Standard Oil controlled 90 percent of the world's oil-refining capacity. Rockefeller was a devout Baptist and represented a liberal Protestantism that valued hard work, stewardship, and good order.[4] He sought to control the oil industry and stamp out competition through tight regulations and policies that only behemoths like his company could navigate.

In contrast, the oil industry of the nineteenth and twentieth centuries was also populated by independent operators called wildcats. Beat out by the Rockefellers in capturing the oil fields of the American East, they expanded aggressively west. Contrary to the establishment, liberal Protestant ethos of the Rockefellers, wildcatters were fiercely independent and disproportionately fundamentalist evangelical. And fundamental to their survival was a free, largely unregulated energy market.

Despite their differences, these two blocs often made common cause in pursuit of pro-oil policies, increasingly within the Republican Party. Yet, their basic oppositional orientation, both in business and in faith, always precluded true harmony. The camps would do business and spiritual battle for much of the twentieth century, competing for new oil fields abroad while fighting for the soul of America back home. In 1979, when Republican candidate for president Ronald Reagan seized on the energy crisis gripping the nation—and on President Jimmy Carter's widely panned response to it—the evangelical wildcatters had an opening. "We must remove government obstacles to energy production," declared Reagan.[5] Wildcat oilmen rejoiced.

Reagan would close the deal at the Southern Baptist Convention's National Affairs Briefing a few months before the election. Addressing a group of hundreds of evangelical pastors and leaders, many of whom made up the backbone of the emerging religious right, Reagan uttered his now famous line, "I know you can't endorse me, but . . . I want you to know that I endorse you and what you are doing."[6] Reagan's platform of aggressive deregulation and a return to traditional family values was music to the ears both of evangelicals exhausted by the blitzkrieg of social change all around them and of the wildcat evangelicals itching to pursue their oil-soaked fortunes unencumbered.

The coalition of evangelicals that coalesced around Ronald Reagan and would go on to become the religious right was motivated not only by abortion or race. They were motivated by both and more—including a wildcat-shaped aversion to any laws or regulations that made it harder to squeeze every last ounce of God-given resources out of the earth. As the religious right moved from the margins of the Republican Party toward its center over the coming decades, this amalgam of commitments went with it and would go on to form the political imaginations of millions of Christians for decades to come.

Why have so many US Christians opted out of climate change action for the last forty years? One reason is because the political story that has formed the beliefs of generations of US Christians regarding how they are allowed to engage in the public square for the sake of the common good—the story written by wildcat oilmen and the religious right and that was enacted by my friends and me in our elementary school mock election—told them that addressing climate change was politically unimportant. Much more pressing were abortion, religious freedom, and the preservation and advancement of deregulated free-market capitalism.

I'll fly away. Most of us have heard it before. Many of us grew up breathing the message in alongside the air in our lungs: *this world is not our home. The Earth is destined for destruction, and we are destined for heaven. We're merely passing through.*

Denial of the spiritual importance of our earthly existence comes wrapped in all kinds of packages. Some of us heard it in the language of premillennial dispensationalism, with speculative timetabling about the decisive end of the world, rapture anxiety, and the full Left Behind series dog-eared on our bedroom bookshelf. Some heard it in the warnings we heard in youth group or Sunday school about the idolatry of pantheism, for a deep love of created things was enough to put any of us in the danger zone of confusing the creation for the Creator.[7] Some imbibed it in church and community cultures so committed to the project of free-market capitalism that the Earth was reduced to little more than property to be bought, sold, owned, and discarded. However we received it, the message was plain: we don't need to care about the Earth because the Earth has no eternal significance.

There has always been a tendency throughout church history to separate the physical from the spiritual, and to elevate the latter over the former. The ancient Christian heresy of Gnosticism was the early church's most formidable. It taught that salvation was found not through the redemptive work of Christ on the cross but through initiation into the secret, spiritual knowledge that Christ came to disclose. The physical world was entirely immaterial.

Various christological heresies flowed from these Gnostic headwaters, and they flourished in the first several centuries of the life of the church. Almost all of them were focused on the dual natures of Christ and sought workarounds to the ancient church's insistence on the radical union of the spiritual and the physical in the incarnation. Arianism, Docetism, Nestorianism, and Apollinarianism all taught in their own way that Christ's humanity and divinity—the material and the spiritual— were not, ultimately, joined in the incarnate Christ.

This stumbling block that plagued the early church, and that plagues it still, is influenced by the Greek Platonic philosophy of dualism, the concept that the material and the spiritual are two separate realities. The spiritual realm is the true, eternal reality while the material realm is but a pale imitation and is quickly passing away.

This dualistic understanding of reality is readily recognizable in much of the church still today, especially as it relates to climate change. Meeting the material needs of those suffering the death-dealing impacts of climate change is eschewed in the name of saving souls instead, as if souls are somehow separable from the physical bodies in which they are enfleshed. A hyperfocus on going to heaven is emphasized as the end of the Christian life rather than a focus on the resurrection of the body and the life everlasting in the new heavens and new earth. And a pronounced fatalism about the need to do anything about climate change pervades much of the church because, after all, the earth is passing away.

Humans have not always conceived of reality in this way, including Christians. Those Platonic heresies were rejected as heresies by the historic Catholic church precisely because they denied the radical integration of physical and spiritual reality, evinced and affirmed most strongly in Christ himself. They missed a key point of the incarnation—namely, that matter matters to God.

When Christians do separate the physical from the spiritual, we repeat the error of these early Christian heretics. We read passages like 2 Peter 3:10-12 in isolation, taking them as proof that the earth will be destroyed by fire, and miss the overwhelming witness of the rest of Scripture that God has big plans for this world. (Hint: it involves joining heaven and earth—the spiritual and the physical—once and for all.) We can recite the Apostles' Creed while being entirely asleep to the radically embodied language it uses to describe the saving work of Christ. We can write and sing songs like "I'll Fly Away" and "I Wish We'd All Been Ready," training our sights on a disembodied, spiritual eternity and miss that the ultimate hope of the Christian story is not heaven at all but the resurrection of the body and the new creation[8]—what New Testament scholar N. T. Wright calls "life after life after death."[9]

And because what we believe about the end of the story shapes the way we play our part in it here and now, we convince ourselves that addressing climate change is ultimately beyond the scope of Christian concern. We are able to say, as prominent evangelical pastor John MacArthur did in a 2008 sermon, "God intended us to use this planet—to fill this planet for the benefit of man. [It] never was intended to be a permanent planet. It is a disposable planet. Christians ought to know that."[10]

Why have so many US Christians been absent from climate change action for the last forty years? Another reason is because the theological story—a heretical story about the separation of heaven and earth, the physical and the spiritual—has shaped their understanding about the nature of reality, the goal of the Christian life, and their ethical obligations in the here and now in ways that make climate change theologically irrelevant.

Big Tobacco, climate disinformation, and the church. In the 1950s, as scientific evidence was piling up proving that smoking caused cancer and other adverse health outcomes, the tobacco industry devised a plan to protect their revenues by waging war on reality itself. The objective was simple: confuse the American public. As an internal tobacco corporation memo put it in 1969, "Doubt is our product."[11]

By all accounts, it was a wild success. Decades passed. The public conversation was mired in disinformation, confusion, and doubt. Tobacco profits held steady. Millions died.

Most have heard at least the broad brushstrokes of this history. And many have heard the sequel: that the fossil fuel industry—inspired by the tobacco industry's success—cribbed tobacco's playbook to sow doubt about climate change. While the popular narrative is linear, recent research seems to suggest that these and other concerted disinformation campaigns were spawned at more or less the same time and conducted by many of the same high-level scientists.[12] In their book *Merchants of Doubt*, historians Naomi Oreskes and Erik Conway argue that these intentional campaigns of obfuscation and misinformation around tobacco, global warming, smog, DDT, acid rain, and more involved many of the same major players and were carried out more or less contemporaneously.[13]

In other words, the last seventy years have seen multiple, sustained, overlapping efforts to intentionally mislead the American public about a variety of issues critical to our health and safety. And study after study has shown that as misinformation metastasizes at an ever-alarming rate in the modern world, one group is more susceptible to these campaigns than any other: White evangelicals.

A 2021 survey from the Public Religion Research Institute (PRRI) found that 62 percent of White evangelicals believed the conspiracy theory that the government is not telling the American people about other treatments for Covid-19 that are just as effective as the vaccine. No other religious group cracked 50 percent.[14] Another 2021 PRRI poll found that 60 percent of White evangelicals continue to believe, without evidence, the Big Lie that the 2020 election was stolen from Donald Trump, again making them the only religious demographic with a majority to do so.[15] According to two separate PRRI surveys, White evangelicals are the subgroup most likely to believe in the QAnon conspiracy.[16] And a 2015 survey from the Pew Research Center identified White evangelicals as the group least likely to agree that climate change is primarily caused by human activity (28 percent) and the group most likely to say that there is no solid evidence that the Earth is warming at all (37 percent).[17]

These numbers don't magically appear out of nowhere. They are a combination of multiple powerful factors, including the political, social, and cultural reasons why White evangelicals are vulnerable to misinformation in the first place as well as millions and millions of dollars' worth of dark money being poured into concerted campaigns targeting conservative, Christian audiences.

There is, at least, some good news on the climate front. A more recent 2020 poll from Climate Nexus, the Yale Program on Climate Change Communication, and the George Mason University Center for Climate Change Communication found that this number has shifted upward. When asked if climate change was caused mostly by human activity, 44 percent of White evangelicals responded in the affirmative. However, this still represented the lowest rate among other faith groups.[18]

Regardless of the very important, but very recent, progress being made on climate change, it says much that evangelicals are among the most susceptible in American society to misinformation. For our purposes, it says that for many of us growing up in evangelical communities, mainstream science about climate change was viewed with skepticism, and important truths about the world in which we live came covered in a thick coating of doubt, confusion, and manipulation.

Why have so many US Christians been absent from climate change action for the last forty years? Another reason is because the sociocultural story of misinformation and intentional obfuscation that has mapped out the basic contours of reality and defined the bright red line between truth and lies has told them that any truth that may exist about climate change is culturally unknowable.

The Consequences of the Stories That Form Us

Of course, there are more stories forming evangelicals' and the broader American church's response to climate change: there is the story that scientific discovery is a threat, rather than a gift, to our faith and that the warnings of climate scientists can't be trusted. There is the story that our comfort and prosperity are God's highest concern for our lives, and that any calls to simplicity or self-sacrifice are against his will. And there is the story, becoming more and more common among younger generations, that the hour is too late and our power too limited to do anything to slow the advance of climate catastrophe.

And all these stories interpenetrate. A basic skepticism of science colludes with misinformation campaigns to make us particularly predisposed to accept climate lies as truth. A dualistic eschatology combines with partisan loyalty to make a commitment to unfettered and extractive free market capitalism not merely a policy preference but a biblical imperative. Sprinkle in the Prayer of Jabez and climate action that requires any kind of material self-denial becomes downright antithetical to Christian living.

Make no mistake: these stories are not neutral. When it comes to addressing climate change with urgency and compassion, the stories we tell ourselves matter immensely, and they have life-or-death consequences for millions of people around the world on the front lines of climate change.

When our political formation is shaped by the story that good Christians vote for one party and one party only, it creates the conditions in which our partisan political identity becomes dangerously conflated with our Christian identity. When this is the case, party orthodoxy can all too easily become Christian orthodoxy. This means that for millions of Christians it is almost impossible to discern whether there might be any daylight between what the Republican Party says about climate change and what their faith might call them to do.

Unfortunately, where any disconnect may exist, their faith is all too often brought into conformity with their politics. A meta-analysis published in the journal *Nature Climate Change* in 2016 found that the most predictive variables for what a given person believes about climate change are values, ideologies, worldviews, and political orientations. Other studies have confirmed this. Our faith is not informing our climate politics; it's the other way around.[19] Since, for so many, their religious identity and their partisan identity are one and the same, any cognitive dissonance that may exist is suppressed or ignored.[20]

Billy Graham foresaw this possibility in the heady days after Reagan's landslide election when the religious right was beginning to come into focus. In a 1981 *Parade* magazine cover story, Graham recounts a conversation he had with Jerry Falwell in which he told him, "It would disturb me if there was a wedding between the religious fundamentalists and the political right." He went on to warn Falwell against the motivations of the New Right, stating that "the hard right has no interest in religion except to manipulate it."[21]

Graham's warnings were prescient but seem to have fallen on deaf ears. Leaders of the religious right like Falwell, Pat Robertson, James Dobson, and even Billy Graham's own son, Franklin, would go on to officiate and bless the very wedding of Christianity and conservative politics that the elder Graham had feared. The consequences four decades later are impossible to miss. Since 2000, White evangelical support for the Republican candidate for president has dipped below 70 percent only once—in the 2000 race between George W. Bush and Al Gore.[22] Yet, even then fifth graders holding their own election implicitly knew that following Jesus in public meant voting Republican.

When Christian identity and partisan political identity become conflated, the moral and political imaginations of generations of Christians are allowed to atrophy. If all we need to do to be good, Christian members of society is vote for candidates with a certain letter behind their names once every four years (or maybe every two years if we're really engaged), there is little need to study, pray through, or discern the nuances of the myriad policy considerations that shape our society—climate and otherwise. The full complexity of political life is flattened to a handful of existential issues. The gradations inherent in even these topics is replaced by a desperate certainty. Gray is traded in for stark black and white.

It's a tempting proposition. Assured by the religious and political leaders we trust that they have done all the hard work for us, we are invited to put our intellectual feet up and put our civic life on autopilot. In a complex world that can all too often feel beyond our understanding and out of our control, the invitation to certainty is a comfort. Yet, it leaves us open to manipulation. Among the more potent object lessons in this kind of vulnerability were the 2016 and 2020 presidential elections, when the overwhelming majority of White evangelical voters supported Donald Trump.

Over the course of the 2016 presidential campaign, it became increasingly clear that Donald Trump was essentially an amoral nihilist. His rhetoric trafficked in grievance and cruelty. Despite his vague and tepid public professions of Christian faith, his life displayed little of the fruit of the Spirit. His worldview seemed to be anchored by delusions of grandeur, and the pursuit of personal power was his guiding light.

And yet, millions of evangelicals flocked to his side because that is what the political story that had formed them told them to do. The right phrases were uttered: "Make America Great Again" (an echo of Reagan's 1980 campaign slogan, "Let's Make America Great Again"), "an end to abortion," and "conservative Supreme Court justices." Christians formed by a story that offers an incomplete vision for Christian citizenship are ripe for manipulation by actors seeking not the kingdom of God but their own power and prestige. Issues that matter deeply to our Christian life and practice, including addressing the climate crisis with urgency and creativity, take a back seat to the ambitions of individual men and women who are interested in little other than their own advancement.

When this political story is combined with a theological story that tells us the world is disposable, with a sociocultural story that tells us the truth about climate change is impossible to suss out among the cacophony of competing voices, and with other stories about the suspect motivations of scientists and the divine blessings of material prosperity, then the Christians formed by these stories have little chance of seeing climate action as an act of faithfulness to God's commands, of neighborly love, or of basic discipleship in the twenty-first century. Instead, climate action is more often than not seen as a partisan threat, a theological heresy, and a dangerous conspiracy—a wild deviation from the stories that have formed them.

Stories form us. And stories are the reason why the US church has been absent from climate action for far too long. So, how do we tell a different story? Luckily, there's always been a different story, and most of us know much of it already. It's just that so many of us have lost the thread. It is the true story of the whole world, for the sake of the whole world. It is the Big Story of God's saving work in the world, creation's central role in the unfolding drama, and our invitation to join in the beautiful, cosmic dance.

3

RECOVERING THE BIG STORY

THE CHURCH HAS BEEN TELLING THE STORY of God's saving action in the world for millennia. Each tradition, denomination, and even congregation recalls it with its own unique accent. Each group may emphasize certain themes, elements, characters, and motifs over others but, unless they are dangerously unorthodox, the basic structure is the same: sin has separated us from God, so God has made a way back to him through the free gift of grace offered through the death and resurrection of Jesus.[1]

In much of the American church, it is entirely possible to hear this story week after week and to hear little, if anything, of the role of the created world in the unfolding drama of God's redemption. If it is present, it is often included as the backdrop for God's action in history—the theater in which the story of salvation is staged, no doubt important but ultimately a passive, inconsequential bystander.

The truth held out to us in Scripture, though, is that creation is much more than a backdrop for the real players and actions. Somehow, in our telling of God's saving work through Jesus, the active, dynamic role of creation in the economy of salvation has been all but lost.

So, I want to take a journey together through Scripture to see if we can recover it, to rediscover how the Bible casts the created world in the Big Story of God's saving action and ongoing mission and to then explore whether this might have anything to say about how we modern Christians should approach something like climate change. This Big

Story held out to us in Scripture is not a disparate collection of proof texts to be harvested in defense of a Christian ethic of climate action. It is a radically coherent love story of a God who creates a good world and entrusts humans with the awesome task to preserve it, whose heart breaks when sin threatens to unravel it all, who does everything he can to bring it back to himself, and who will finally be united with it in perfect, right relationship once more.

God Creates a Good World and Entrusts Humans to Preserve It

Genesis 1: Ruling alongside creation's true king. In the beginning, we see a God who, in his perfect freedom, pours his love outward and makes order out of chaos. We see a God who revels in the goodness of the created world for its own sake well before humans are ever on the scene, stepping back to marvel at the works of his hands and to utter breathlessly, "It is good!"

The Hebrew word for "good" in this passage is *tov*, and Genesis 1 employs it seven times in some variation of the phrase "and God saw that it was good." To the ancient Hebrew mind, numbers communicated more than numerical value. They also conveyed a sense of the cosmic significance of the world and theological truth about God. The number seven carried meaning tied to wholeness, completeness, and even holiness. When Genesis 1 tells us that God calls creation good seven different times, the Scripture is communicating something important about the nature of that goodness. It is total. Creation, as God sees it in Genesis 1, is maximally good. As good as it gets.

And after reveling in the sheer goodness of it all for five glory-filled days, God comes to the penultimate day of his tour de force. And on it, after creating badgers, beavers, and billy goats, God creates one more very peculiar creature: human beings. And though our placement next to all the other land creatures on the sixth day seems to be a subtle reminder of our place within the midst of creation rather than above it, the passage includes these unique words about God's creation of humankind:

"Let us make [human beings] in our image, in our likeness, so that they may rule over the fish of the sea and the birds in the sky, over the livestock and all the wild animals, and over all the creatures that move along the ground."

So God created [human beings] in his own image,
in the image of God he created them;
male and female he created them. (Genesis 1:26-27)

The theological anthropology of Genesis 1 teaches that we humans are both creatures and image bearers. We are embedded in creation, mutually bound up with all the other creatures and equally dependent on the gifts of water and air, protein and calories for our continued existence. Yet, we are unique in the midst of it. Humans alone among God's creatures bear the Creator's image. What's more, God intends to give us alone the authority to rule over all other creatures. Unfortunately, the impulse of the church, right up to today, has all too often been to erase this dynamic tension by privileging our uniqueness over our mundanity—to read into Genesis 1 a divine permission slip to do whatever we see fit with the rest of God's created world in pursuit of human comfort and advancement.

Yet, despite this misuse of our image-bearing capacity, there can be no doubt that Genesis is clear about humanity's uniqueness. Theologians have puzzled over the precise nature of the *imago Dei* for millennia. Is it our capacity for language? Human consciousness and rationality? Our irrepressible social nature that makes us crave community and relationship? Our capacity and drive to create and innovate?

While each of these possible answers to the question is compelling in its own right, each also has its limits. If the *imago Dei* is the capacity for language, then do infants, those who are mute, or the profoundly disabled not bear God's image? If it is the ability for rational thought, young children and the cognitively impaired are again left out in the cold. If it is the innate need for community, as the triune God himself is being-in-communion, then what to make of those who lack meaningful community? If it is our irrepressible creativity, then what to make of the victims of horrific injury left in a vegetative state who will never again put pen to paper or lift their voices in song? Has the image of God somehow left them?

What does it mean for humans to be made in the image of God? Most responses to this question will immediately attempt to identify the particular gifts that make humans unique among the rest of God's creatures. This approach to the question begins from the perspective of the privileges afforded to humans by virtue of their image-bearing status, and always bumps up against limitations. But what if instead of looking for which privileges are unique to humans, we instead looked for what responsibilities are unique to us? What if we asked not, What special privileges does the image of God impart on humans over against other creatures? but rather, What unique responsibility does the image of God call forth in humans toward the rest of creation?

When the *imago Dei* is approached through the lens of responsibility rather than privilege, it begins to unlock Genesis 1 in new and important ways. Take the following two Hebrew words in Genesis 1:26: *tselem* and *demut*. Translated as "image" and "likeness," respectively, these are often taken to be mere synonyms. This is an understandable interpretation for English speakers. After all, the English language boasts the largest vocabulary of any spoken or written language in the world. English has nouns, verbs, and adjectives to burn. And evocative English writing burns them, clumping synonyms together to offer richness but not necessarily deeper meaning.

The Hebrew language can likewise glory in synonymity—particularly Hebrew poetry and prophecy. Yet, Hebrew also has a famously sparse vocabulary, and most Hebrew words contain a wide range of meaning. This means that Hebrew words are often called on to do considerable heavy lifting. While repetition and synonymity do flourish in Hebrew writing, it is not always the only objective. There is often something else going on as well.

When two Hebrew words of similar denotation are used next to each other then, as *tselem* and *demut* are in Genesis 1:26, the purpose can indeed be the simple joy of wordplay and repetition. It can also be to provide additional meaning. *Demut*, for instance, has the smaller range of meaning of the two. It carries connotations of a physical structure or physique, and "likeness" is a good rendering. Humans—in our physical createdness—are somehow made like God, similar to him in a way unique from the rest of creation. The Hebrew word *tselem*, on the other hand, prioritizes a less physical relation. Mirror, mirage, and reflection are all faithful renderings of *tselem*. It also carries more abstract connotations of figurative representation. Interestingly, it is often translated in the Old Testament as "idol."

This may seem strange at first blush. After all, aren't idols big no-noes in the Bible? Yet, what is an idol but the reflected representation of a higher power? In the ancient Near East, well before television or the internet, new rulers and kings of large empires would often erect statues of themselves throughout their territory after their coronation. The purpose was to communicate their power and authority to the subjects of their far-flung provinces. It was a way to announce the coronation of a new ruler, as well as a means whereby subjects might learn to recognize their ruler. In the event of a royal visit, subjects would be able to discern their true king from any charlatans seeking to wrap themselves in the mantle of majesty without any real claim to it. Obviously, the statues of these rulers were not the rulers themselves. They were instead intended to point to the ruler, to remind the subjects who their true king was.

In the pairing of *tselem* and *demut*, we see the wonderful nuance of the *imago Dei*. Humans are uniquely privileged to share in God's likeness, yet we are bound to exercise that privilege in a particular way that reflects the rule and reign of the Creator. Image bearers of God carry not merely a collection of discrete privileges divorced from the rest of creation but an immense and sacred responsibility toward it. The responsibility, through our actions in the midst of creation, is to point the rest of creation to its true and only King, to mirror the joy and delight of creation's Maker by also using our own creative words to speak goodness over all created things, and to represent Creation's true king in the midst of creation. We are not meant to usurp the authority of the true king or to use our status as a license to abuse and exploit. Instead, we are called to live in the midst of creation in such a way as to remind the created world—and ourselves—who the true ruler is.

Genesis 2: To serve and protect creation. By reading Genesis 1 through the lens of responsibility rather than privilege, our proper place in creation begins to come into focus. Genesis 2 makes the focus even sharper.

Biblical scholars have long recognized that Genesis 1:1–2:3 and 2:4-25 are two distinct creation narratives. A careful reading of the chapters makes the differences between the two pretty obvious. No waters cover the earth at the beginning of Genesis 2 as they do in Genesis 1; instead, the earth is dry and barren. Humans are created second to last in Genesis 1 (the last and crowning creation is the Sabbath), after all the plants, trees, birds, fish, and creatures that move along the ground. But in Genesis 2, humans are created before anything else. God's creative acts in Genesis 1 are orderly, sequential, and instantaneous. In Genesis 2, God's acts are deeply intimate and physical, forming creatures with his hands from the dirt of the ground.

Now this isn't because the Bible is fuzzy on the details or somehow at odds with itself about the specifics. It's because the two narratives are communicating different truths about God as Creator. The language and style employed in Genesis 1 is high poetry, the kind of majestic language that could live comfortably in the royal archives alongside official decrees and pronouncements of the king. God creates alone and does so easily, merely speaking his good works into existence, as opposed to the gods of ancient Israel's neighbors who created collaboratively through violence, conquest, and bloodshed. Just as a true king speaks and his word is law, so too does God speak and creation's laws are established.

The central message of Genesis 1 is that God, and only God, creates because God alone is creation's true king.

By contrast, the language of Genesis 2 is folksy, the kind of language you'd hear around a campfire. Rather than high-minded poetry, it's a folktale. Rather than a transcendent God speaking his will into being, God kneels in the mud and works with his hands. The central message of Genesis 2 is that God is passionately and intimately involved with his creation.

These are two stories with two different emphases, but both are true. It might seem strange to have two separate and divergent stories right next to each other. But it's not so strange when we remember that Genesis is much less interested in answering the *how* and the *when* of creation—questions our modern minds are obsessed with—and is much more interested in the *who* and the *why*. And in both accounts, the *who* and the *why* are crystal clear: God is the *who*, and love is the *why*.

Like Genesis 1, Genesis 2 goes out of its way to put humans in their place as creatures in the midst of the rest of the created world. And it does it with a pun. The tragedy of wordplay is that it's prone to getting lost in translation. In English, we read that "the LORD God formed a man from the dust of the ground" (Genesis 2:7). No wordplay there. In the original Hebrew, however, a delightful pun is employed to emphasize the deep connection between the man and the ground from which he's formed. The Hebrew word for man here is *adam*, the source of our English transliteration "Adam." (Interestingly, Hebrew manuscripts of Genesis 2 never name the first man. He is merely "man"— *adam*.) The ground from which the *adam* is formed is the Hebrew word for soil: *adamah*. Human beings are *adam* from the *adamah*. We are, quite literally according to the Hebrew, "soil people." From our very beginning, we are intimately bound up with the rest of creation.

After God forms this creature and breathes his own vivifying breath into it, he plants a garden. Then God takes the soil-man, places him in the midst of the garden, and gives him a command: to *avad* and *shamar* it.

These two Hebrew words are often rendered as something like "to till and to keep [the garden]." But the Hebrew word *avad* is used often throughout the Old Testament, and most of the time it is used in the context of service—even slavery. This is how Joshua uses it in Joshua 24:15 before the Israelites enter the Promised Land, when he proclaims that on the other side of the Jordan, he "and [his] household . . . will serve [*avad*] the LORD" (Joshua 24:15). The second Hebrew word, *shamar*, is also used throughout the Old Testament, and it means to actively guard, to proactively and preemptively protect from harm. A famous example of a passage that uses the word *shamar* is Psalm 121:

> I lift my eyes to the mountains—
>   where does my help come from?
> My help comes from the LORD,
>   the Maker of heaven and earth. (Psalm 121:1-2)

The Jewish people would sing this psalm on their way up to Jerusalem for the high festivals, a journey that would take many of them along dangerous and precarious routes. Psalm 121 uses the word *shamar* six times in eight verses to describe a God who so actively and jealously protects his people from harm that their feet can't even slip on a loose rock on the road up to Jerusalem without God catching them. This is the kind of jealous, active protection described by the word *shamar*.

When these three translations of *adam/adamah*, *avad*, and *shamar* are taken together, Genesis 2 can perhaps more accurately be rendered as "The LORD God took the soil-man and placed him in the garden to serve and to protect it" (Genesis 2:15).

All of this matters, not just as an intellectual exercise but because the way we translate Genesis 2 changes the shape of the command. Rather than a disconnected custodian maintaining creation—or worse, a superior extracting value from an instrumentalized inferior—the human beings in Genesis 2 are envisioned as embedded members of the created order imbued with a unique, God-shaped responsibility to serve and protect the rest of creation, as they are served by creation in return. It matters because this command comes on the heels of Scripture's first God-given order to his image bearers in Genesis 1 when he tells them to subdue and rule over creation. When the command to rule is placed next to the command to serve and protect, it becomes clear that this is one, coherent command: rule through service.

When Genesis 2 is allowed to sit next to Genesis 1 as a couplet of instruction, the proper shape of our relationship to the rest of creation comes into sharper relief. This is what it means to rule over the fish in the sea and the birds of the air and the creatures that move along the ground: to be in a special relationship of service with the earth and all its creatures. We are expected to actively and jealously protect it with the same tenacity that God protects his people. We are commanded to rule over creation in the same way that creation's true king rules, a king who, finding himself in the midst of creation,

> did not consider equality with God something to be used to
>   his own advantage;
> rather, he made himself nothing
>   by taking the very nature of a servant, . . .
>   he humbled himself
>   by becoming obedient to death. (Philippians 2:6-8)

Jesus does not rule through dominance, extraction, or exploitation, but through humble service, sacrifice, and by seeking the good of that which is ruled. He rules through service.

And if rulership through service sounds like an oxymoron, perhaps our conceptions of power and authority have been shaped more by the prevailing political consensus than by Jesus. Jesus—creation's true king from whom we derive our own authority to rule—made a habit of exercising his authority over creation through humiliating incarnation, humble footwashing, and sacrificial death. Our authority to rule creation derives only from him. We only ever exercise our authority alongside him. If our dominion of creation doesn't look like Jesus, then we're doing it wrong.

For too long, we humans have been doing it wrong. We have usurped Christ's throne and fancied ourselves kings of the universe, and the created world our footstool. Yet, Genesis 1 and 2 hold out to us a different way: sacred responsibility, not selfish privilege; humble service, not domination.

God's Broken Heart

Of course, it doesn't take long for the image-bearing soil-creatures to overstep their creaturely bounds and jeopardize everything. When sin enters the good world God has created, its effects permeate every corner. When God is describing the consequences of human disobedience in Genesis 3, he underscores how deep the break goes by declaring that the *adamah* will no longer willingly collaborate with the ministrations of the *adam*. Instead, humans will struggle and toil to bring forth food for themselves (Genesis 3:17-19). When Cain commits the first murder only one chapter later, God will call the *adamah* as a witness against him, proclaiming that Abel's blood is crying out to him from the ground where Abel lay slain (Genesis 4:10).

And yet, God refuses to abandon any of it. Mere chapters later, when the primordial waters once again cover the face of the earth and Noah's life capsule floats lonely on the waters, God will make a new covenant not only with his human image bearers but with the earth and with all living creatures on the earth (Genesis 9:1-17). After sin enters the world, God makes it clear from the very start: he intends to rescue all of his beloved creation.

Job 38–41: A love letter from the whirlwind. Why does God make a covenant with Noah, with all living creatures, and with the earth itself after the flood? He does it because it is abundantly clear that he is recklessly, hopelessly in love with all of it. After all, God doesn't try to hide it. It's splashed across the pages of Scripture. His love for all created things is demonstrated in the commands in Leviticus to provide not only people a sabbath but the land as well and to never forget that the land belongs to God (Leviticus 25:4, 23, respectively); in the myriad psalms that attest to God's intimate care, concern, and compassion for all creation (Psalms 8, 19, 24, 65, 104, and more); and in the incarnation itself (more on that in a bit).

And in four chapters at the end of the book of Job, God pens perhaps his most moving love letter in Scripture to the world that he adores. Job 38 comes on the heels of thirty-seven chapters of complaint and despair. Job, a righteous man by all accounts, has had everything possible ripped away from him over the course of cascading tragedies. His home is destroyed, his children are killed, and even his body is given over to painful sores and disease. In the tradition of Jewish covenant theology, Job calls God to account. By Job's reckoning, he has lived up to his end of the covenant bargain. He has worshiped only God, he has tithed the gifts he has received, and he has been a righteous man. God, on the other hand, seems to be asleep at the wheel. How else to explain Job's suffering?

So, Job has a go at God. He demands God give an account for his predicament. For thirty-seven chapters, God lets Job run with the line. And then the whirlwind breaks open: God's turn.

There can be no doubt that Job's encounter with God in these chapters is intense. After all, God is described as speaking out of a storm. God's monologue is shot through with zingers and more than a little sarcasm. The basic structure of God's response to Job is a long series of rhetorical questions intended to diminish Job's self-importance with every question he clearly cannot answer.

Yet, the passage is not primarily a dressing down. Nor can it be rightly described as a legal treatise outlining all the ways in which it is within God's rights to allow Job to suffer as he has. When read straight through, Job 38–41 has the prevailing feel of a love letter—a love letter written to God's creation. It positively drips with affection. The profound intimacy with which God interacts with his creation is a gift to behold.

God teaches the dawn its place (Job 38:12) and takes both darkness and light to their dwelling places (Job 38:20). He raises his voice to the clouds and sends lightning bolts on their way (Job 38:34-35). He hunts the prey of the lioness and provides food for the raven (Job 38:39-41).

Yet, this is no transactional business deal between God and his creatures. This is intimate, loving relationship. Job 39 tells us that God knows when the mountain goat gives birth and that he counts the months until the doe bears her fawn (Job 39:1-2)—can't you just picture God checking off days on a calendar, eagerly anticipating the baby deer's arrival with bated breath? God notices the laughter of the wild donkey (Job 39:7), the joy of the ostrich (Job 39:13), and the pride of the horse (Job 39:19-25), not unlike the way a parent notes the unique personalities of her different children. God dedicates half of Job 40 to his pride in the strength of Behemoth (possibly the hippopotamus) and all of Job 41 to the grace and wildness of Leviathan—a mythical sea monster that God seems to have created simply for the sheer joy of watching it sport in the waves. This is not the language of a distant, uninterested God. This is the language of a God who is intimately involved and passionately in love with his created things.

It's as if God's response to Job is to say, "If it feels to you like I've nodded off on the job, check my references. I've been feeding lions, midwifing fawns, laughing with Leviathan, and directing the rain and snow across the world." Indigenous theologian and wisdom keeper Randy Woodley is fond of reminding us that every moment of every day God is in relationship with parts of creation that are completely outside of human perception.[2] Such is the scope of God's great love for creation, and the limits of our own experience of it.

Why, then, wouldn't God's breaking heart extend to all of creation, and his rescue mission not include it all?

God Does Whatever It Takes to Bring All of It Back to Himself

John 1: God binds himself to creation. If you're like me, you've heard the Christmas story now for decades. It's easy, then, to go to sleep to the jaw-dropping scandal of God-made-flesh. If we can approach John 1 with new ears and fresh eyes, John will show us the depth of God's commitment to all of his creation.

When we read John 1 with intention, we notice pretty quickly that it sounds a lot like Genesis 1. The author even starts with the same words: "in the beginning." This is no accident. In fact, it is precisely part of John's point. This Jesus who lived and breathed and walked among humans, who taught and slept and lost his temper—this Jesus was not merely a significant teacher or an extrapious individual. He is the very capital O One who hovered over the darkness and made order out of chaos that first day. He is the very capital O One through whom all things came to be—the very God who created the good world in which he takes such delight.

To drive this point home, John uses a peculiar Greek word to describe Jesus: *logos*, translated most often as "the Word." This can strike our modern ears strangely. For those of us who grew up in the church, it's likely we've simply become so used to this metaphor that we no longer puzzle at its meaning. Yet, in the context of the two cultural worldviews that shaped the authors and first readers of the New Testament—Hebrew and Greco-Roman—this choice made all the sense in the world.

In Greco-Roman cosmology, the *logos* was understood to be the animating force that infused the world with being and life. The *logos* was the ever-present energy—variously described by the competing philosophical schools as wisdom, logic, rationality, or beauty—that both created the world in the beginning and sustained it from one moment to the next.

For its part, Hebrew cosmology was shaped by its understanding of words. In the Hebrew imagination, words were not simply linguistic tools of communication. Words were the building blocks of reality. It is no mistake that the Hebrew phrase *vayyomer* ("and he said") is found eleven times in Genesis 1. The idea that God created the world with words was no flashy metaphor. It reflected a fundamental understanding of how the world works. God created the world with words, and God's words of love and covenant faithfulness continue to create the world from one moment to the next. In fact, the Hebrew word *davar* can be translated as both "word" and "action"—so closely tied is the Hebrew understanding of speaking and doing.

This actually makes a lot of sense. Think about a time when you were encouraged by the kind words of a dear friend in a period of vulnerability or self-doubt. Or remember a moment of frustration when you spat out words of hurt toward another that you desperately wanted to take back. Our words go out into the world, creating realities of trust and support or of pain and betrayal for ourselves and those around us.

It's clear that John is going for something specific by framing the incarnation of Jesus in this way. He could have articulated the mind-bending paradox of the incarnation of the eternal Son of God in myriad ways, but he chooses to tap deep into the cosmology and metaphysics of the Greco-Roman and Hebrew worldviews, to the very nature of reality and creation. Why would John find it so important to tie creation and the incarnation together so tightly? Perhaps it is because in the incarnation we see God's culminating affirmation of created things.

The idea that God would assume physical existence—that he would wrap himself in flesh and bone, tendon and cartilage—was an insurmountable obstacle for many early Christians. The formidable Gnostic heresy, which we discussed in chapter two, is proof of just how hard it was for the early church to accept this paradox. Gnostics simply could not make sense of a perfect God assuming fallen flesh. As we saw in chapter two, it's a stumbling block that has continued to dog the church and that lives on in various forms today with devastating consequences.

But the incarnation should come as no surprise to those who have been paying close attention to the Big Story of God's loving action in the world. The Gospel writer has been paying attention and signals as much in that favorite Advent verse: "The Word [*logos*] became flesh and made his dwelling among us" (John 1:14).

The Greek word translated as "made his dwelling" is actually more specific than that. It literally means "tabernacled." For the Hebrews wandering around the desert, the tabernacle was God's manifest presence among them. When it was erected, God was literally inside it. And if you wade through all the meticulous instructions that God gives the Israelites for constructing the tabernacle, you might be stunned by what you find in Exodus 26:14: "Make for the tent a covering of ram skins dyed red, and over that a covering of the other durable leather."

God instructed the Israelites to put a layer of skin on the outside of the tabernacle—to literally wrap his presence on Earth in skin. The Hebrew readers of John 1:14 were used to hearing that the "Word became flesh" because, ever since the building of the tabernacle, the Creator has been putting on skin to be near his creation.

The God who created all things, says John, loves his entire world so much that the thought of losing it to the power of sin and evil was simply unbearable. Rather than consign it to death and decay, God chose to decisively enter into the midst of creation in order to redeem it from the inside. And he chose to do so by taking on the very stuff of the created world: the flesh and blood of his own image bearers. Rather than the antimatter, antiphysical dualism of Gnosticism, the incarnational theology of John 1 is unequivocally creation affirming.

Scripture is clear that matter has always mattered to God. God creates the world—sun, rocks, birds, bacteria—and revels in its goodness. He creates humans and calls us to be co-protectors of his beloved world. He lavishly displays his love in the world around us and is always using the stuff of the world—be it flesh, water, bread, or wine—to reveal himself. Nowhere is this more powerfully on display than in the incarnation, when the Word that created matter assumes it in order to redeem it.[3]

And even more, the resurrected Christ takes human flesh up into heaven. It is safe to assume that, since the disciples watched Jesus ascend until a cloud "hid him from their sight" (Acts 1:9), Jesus was in a visible and recognizable form when he journeyed from our space back to God's space. It is also logical to conclude, as orthodox Christology has assumed for millennia, that Christ remains incarnate in heaven to this day. I'll never forget the first time my college religion professor asked, "Does

Jesus still have a body?" It completely blew my mind, and no wonder. Consider the ramifications of such a Christology! Atoms and molecules have been forever woven into the eternal life of the Godhead. Our human form is in the presence of God right now, seated at the right hand of the Father. The stuff of creation is in God's space, forever, all because of God's deep love for the world that he created. I can't imagine a stronger affirmation of the goodness and inherent dignity of the stuff of creation than for the Creator to take it on, making it a part of himself for eternity, in order to rescue it forever.

Colossians 1:15-20: What if Paul meant "all things"? In Colossians 1, Paul takes pains to spell all of this out for his audience of fresh followers of the Jesus Way in Colossae. He seems to understand that before he can encourage them in their freedom from the Jewish Law or exhort them toward proper living, as he will in coming chapters, the foundation must be poured. Paul does most of this work in Colossians 1:15-20.

In form, it is a poetic hymn about the Christ, and it is some seriously theologically dense writing. It would take chapters and chapters to unpack all that Paul is saying in these few verses—chapters that have already been written by biblical scholars more insightful and immersed in the text than I. For our purposes, I simply want to highlight a curious characteristic of Colossians 1:15-20, introduced to me first by the biblical scholar and ethicist Steven Bouma-Prediger,[4] and to suggest that this, too, is evidence of God's abiding love for all he has created and of his intention to bring it all back to himself.

In Paul's six-verse magnum opus on the nature, character, and work of Christ, he uses a tiny, two-word phrase more often than any other: *ta panta*. It is the phrase in Colossians 1:15-20 that is translated into English as "all things."

Upon a reread, the phrase "all things" jumps out everywhere: Christ created all things, and all things were created for him (Colossians 1:16). Christ is before all things, and in him all things hold together (Colossians 1:17). Christ is first in all things (Colossians 1:18). And God is reconciling all things through Christ (Colossians 1:20).

The phrase shows up six times in the English translation of Colossians 1:15-20, and some form of the root *pas* ("all") is found eight times in the original Greek. If all my years in Sunday school, Christian day school, Christian higher ed, and seminary have taught me anything, it's that if the Bible repeats itself, pay attention.

This is especially true of texts that were intended to be read aloud, as the New Testament Epistles undoubtedly were. Given that most people in Colossae (and Ephesus, Thessaloniki, Philippi, Galatia, Corinth, and Rome) could not read or write, these letters were read aloud by a scribe for the entire gathered assembly. When writing for the ear, repetition becomes a crucial aural device to emphasize central rhetorical points. It signals to the listener, who is only human after all and may be distracted by a buzzing fly or a grumbling stomach, that this point is especially important. Repetition increases the chances of penetrating the preoccupied human mind and burrowing into the guarded human heart. If Paul wanted his listeners to hear one thing only, it's reasonable to assume that he would repeat it more than any other point. *Ta panta* is repeated more than any other word or phrase in the Colossians Christ hymn.

What could this repetition of the phrase "all things" mean for what Paul was trying to tell his original audience? For what he is trying to tell us today? At the very least, it holds out to us the possibility that while the saving work of Jesus on the cross is certainly meant for humans, God very well might have much more than *only* humans in God's saving sights.

We Christians can be very good at convincing ourselves that the Bible doesn't always really mean what it says—especially (and perhaps scandalously) with Jesus himself. Jesus didn't *really* mean we should love our enemies and turn the other cheek rather than retaliate when provoked (Matthew 5:38-40). He didn't *really* want us to sell everything we own and give to the poor before we try to follow after him (Luke 18:18-29). The hard truths and inconvenient directives are transformed into metaphor and symbol, insulating us from onerous ethical demands and protecting our theological systems from critiques that might cause us existential heartburn.

I think we do this with Paul sometimes too, and especially in this passage. "All things" is so often taken to be a clever literary flourish, a poetic synecdoche whose true parameters extend no further than the human heart. But what if we're wrong?

What if Paul means exactly what he says? What if he means that precisely because all things were created through Christ and for Christ, all things are also being reconciled back to Christ through his blood shed on the cross? And by "all things," what if he means human hearts and Redwood forests, diving dolphins and soaring seagulls, impenetrable mountain summits and unknowable ocean depths? After all, if God so loved the cosmos that he sent his only Son to save it (John 3:16), why wouldn't the whole of the cosmos be included in Christ's redeeming work?

For some millennial Christians like me (and Generations X and Z for that matter), this could be a difficult mystery to accept. After all, many of us grew up in a hyperindividualistic Christian subculture obsessed with personal salvation. Altar calls, church camps, purity lectures, and Christian music festivals all delivered the same laser-focused message: Christ died for *you*; *you* have the choice to accept salvation and live for him. And what did "living for him" mean? Usually, it was a moralized life marked by private habits (like personal devotions and prayer) and public piety (like abstaining from sex, drugs, and alcohol). The focus was almost exclusively on our personal relationship with Jesus, the implications of which usually penetrated no further than our own hearts or our habits when we were mostly alone. When Jesus did make demands on our public and social life, it was usually to invite our lapsed Catholic coworker to church or to be ready to pray the Sinner's Prayer with a seatmate on a cross-country flight. It was a distinctly American faith—an individualistic, soul-focused, this-world-is-gonna-burn, grab-your-get-out-of-hell-free-card-while-you-can kind of Christianity.

To this kind of faith, Paul offers the invitation to consider that if a theology of salvation doesn't extend beyond the human soul, then it might be incomplete. Perhaps the goal of Christ's death and resurrection extends far beyond Paul, and me, and you. The scope of God's saving purposes in Christ, in fact, might just have the entire created world in view—all things.

The inclusivity of Paul's soteriology in Colossians can feel shocking, but it's also exhilarating. God, it appears, is so in love with his creation that he refuses to consign any of it to the destruction and decay wrought by sin. Instead, he intends to bring *all* of it back into right relationship with him through the redeeming work of the firstborn over all creation. While the saving work of Christ was accomplished for me and for you, it also seems to have been done for much more than *only* you and me.[5]

God's Big Plan for Creation

Revelation 21:1-5: Everything (re)new! If Paul makes the case that God's reckless love for creation will drive him to snatch all of it from sin's gaping maw, then John the Revelator offers an eschatological vision to confirm it in Revelation 21. John, consigned to a life of exile on the rocky outcrop of Patmos in the Aegean Sea, is acquainted with suffering. His audience of first-century Jews was too. Most scholars date the writing of Revelation to sometime around 94–95 CE. This means he is writing at the tail end of a deeply tumultuous and traumatic period in Jewish history.

God's chosen people had a precarious grip on self-determination from the moment they settled into the Promised Land. Ancient Israel was perpetually surrounded by the major power players of the day—be they Assyria, Babylon, Persia, Greece, or Rome. Given their location at the crossroads of the major trading routes of the ancient Near East, they often found themselves buffeted by the geopolitical actions of their much more powerful neighbors. The Northern Kingdom of Israel was eventually hauled off into exile in Assyria in 722 BCE, leaving only two of the original twelve tribes of Israel intact. These tribes would follow their kindred shortly after when they were forcibly deported from their homes by Babylon in 587 BCE. While the captives from the Assyrian deportation would be scattered to the wind and mostly lost to history, the Babylonian exiles would eventually return to Jerusalem in 539 after Babylon was defeated by Persia.

The next five hundred years or so provided a period of relative geopolitical calm that allowed the Jewish people to tend to their institutions and govern themselves in peace, albeit with some serious harassment from Greek invaders. That all changed, though, with the rise of the next major empire: Rome. As the Roman army set about gobbling up territory and subjugating neighboring populations, Jerusalem once again fell in 63 BCE.

From the start, Roman occupation of Jerusalem and Palestine didn't go well. Despite its best efforts, Rome was never able to fully snuff out the simmering resentment and bubbling rage of the Jewish people at their overtaxation and lack of self-determination. For more than a century, this resentment and anger would slowly build into a nationalistic fervor that began to expect God's promised salvation to come in the form of a political deliverance from Roman occupation. It was this nationalistic expectation that drove the crowds greeting Jesus upon his entry into Jerusalem for the Passover (itself a remembrance of God's decisive deliverance from Egypt, another oppressor) with palm branches and shouts of "Hosanna" (palm branches serving as a national symbol of the Jewish people[6] and *Hosanna* meaning "save" or "deliver" in Hebrew; see Matthew 21, Mark 11, and John 12). It was this nationalistic fervor that caused the ruling Jewish leaders and Roman authorities at the time to feel deeply threatened by such a dangerous political demonstration and to seek to take the troublemaking itinerant preacher off the board for good.

Even after Jesus' crucifixion and resurrection, the Jewish people in Palestine living under Roman occupation continued to chafe. The climax of first-century Jewish nationalistic ambition came in 66 CE when the various revolutionary factions joined together in a united coalition to expel Roman occupying forces from Jerusalem. They succeeded in pushing the Romans out of the city, and much of the rest of the surrounding countryside, and established a revolutionary governing authority that ruled a free Jewish people until Rome was able to re-double its forces and launch a fresh attack on the revolutionaries in 70 CE. They were crushed by the superior military might of the Romans. Jerusalem was sacked, the temple was burned to the ground, and the nascent Jewish state collapsed.

Twenty-five years later, John—a member of the Jewish diaspora— puts pen to paper. He is trying to capture a mysterious set of visions he has received and to interpret them for a scattered people still living under the shadow of trauma and despair, a people shaped by centuries of subjugation who, only a generation ago, lived through the destruction of their most sacred religious site, the locus of their communal identity and the very dwelling place of God on earth. The trauma of the temple's destruction is impossible to overstate. To the Jewish people at the time, it was reminiscent of the ancient captivities in Assyria and Babylon. It recalled to their memory the prophecies of Ezekiel, when God picked up and left the temple (Ezekiel 10), leaving his disobedient people to their fate at the hands of much more powerful enemies. God, it ap-peared, had abandoned them again. The smoldering ruins of the temple were proof positive of that.

Into this trauma and despair, John speaks a new word from God and gives his first-century Jewish audience suffering under Roman rule a lifeline. "Though you are suffering now," says John, "just look at what God has in store!"

And what did God have in store for his people? What exactly is the hope that John holds out, both to the first-century Jewish community and to us today? The hope is this: that God's final, culminating act of salvation is to come back—this time for good. John tells the community that expectations of a political victory over Rome were misplaced be-cause God intends to do away with the idea of Rome (and Assyria and Babylon) altogether. Oppression and violence will be impossible be-cause God intends to join heaven and earth—God's space and our space—once and for all. And he will do so not by sucking up disem-bodied human souls into an ethereal heaven where we will float on clouds, sprout wings, and play harps (which sounds a lot more like Greek philosophy than Jewish theology, a lot more Plato than John the Revelator). We will not go to God. Instead, God will come to us, just as he's been doing from the very beginning—just as he did in the garden, in a burning bush, in a pillar of cloud and fire, in the tabernacle, in the temple, in Jesus.

We know this because of what John sees: a new Jerusalem coming down out of heaven to earth. Jerusalem had been the city that gave his Jewish audience meaning and identity, the city that held their hopes and dreams. Now this city lay in ruins yet again at the hands of a conquering force directed by a ruler other than Yahweh. And as the city was ruined, so was God's holy temple, the dwelling place of God's very presence on earth. Rather than being rebuilt by human hands yet again, as it had been after the Babylonian exile and thereby was left vulnerable to future attack, John assures the Jewish diaspora that God plans to rebuild Jerusalem himself. By seeing the new Jerusalem come down out of heaven to earth, John is effectively saying, "God's presence will come back, and this time he will never leave us again."

But what about that word *new*? John uses the word to describe God's coming future, featuring a "new heaven," a "new earth," a "new Jerusalem," and finally, putting these triumphant words in the mouth of the one seated on the throne, "I am making everything new!" (Revelation 21:5). Does this mean, then, that this world is, in fact, passing away and that God will start over? Does it mean that John MacArthur is right and that the current creation has no eternal destiny?[7] Do we live in a disposable world, and should we therefore feel fine about throwing it away? In a word: no—at least, not if we take the Greek of Revelation 21 seriously.

In the Koine Greek in which Revelation and the rest of the New Testament is written, there are two words that are rendered in English as "new": *neos* and *kainos*. *Neos* refers to something that is brand new. When a new home is built on a piece of property, this house can be said to be *neos* because it never existed before it was built. We English speakers understand the meaning of *neos*.

Kainos, on the other hand, refers to something that is new in form, substance, or quality. It carries connotations of renewal and restoration, of taking that which is and bringing it to its full, intended purpose and potential. The house that was once *neos* when it was built can never be *neos* again. But after a complete remodel, it can become *kainos*—refreshed and brought back to its fullest intention once more. The tired and satisfied homeowners can look on their handiwork and say of their home, "Like new!"

Paul uses *kainos* in 2 Corinthians to describe the nature of a new believer in Christ: "Therefore, if anyone is in Christ, the new [*kainos*] creation has come: The old has gone, the new [*kainos*] is here!" (2 Corinthians 5:17). This makes sense since no human once born can be born a second time (just ask Nicodemus, John 3:1-21). Instead, through union with Christ, they become renewed.

In Revelation 21, John uses *kainos* exclusively. John, like Paul, is making plain that God's not starting over with his people or with his creation. Instead, God is restoring his masterpiece. In other words, God is in the business of making all things new, not making all new things.

This means, then, that all things have a destiny, a place in God's coming good future. And by living like we believe this truth, we participate in and bear witness to this coming age, however imperfectly. By working to stop pollution, to preserve biodiversity, and to slow climate change, we join our voices with that emanating from the throne. We proclaim that this world matters, that God has a good future for it, and that even now God is making everything *kainos* new!

In this one small Greek word, there is tremendous hope, not just for the Jewish diaspora of the first century CE but for us today. Our world is beset by crises: economic, public health, democratic, ecological, racial, mental health. The list seems truly endless. John's original audience knew crisis too. Yet, the beautiful truth that he held out to them—and that he holds out to us today—is that even though the headlines scream otherwise, the future of God's world is secure, and his purposes will be accomplished.

But if we think that this means we're off the hook and free to put our feet up until God's ultimate will is fully realized, we have another think coming. Jesus was clear what our task is as we await God's coming good future, and Paul and other New Testament writers make this clear too. Our task as we wait for heaven and earth to be united in perfect justice, mercy, and peace is to live now—by the power of the Holy Spirit—as if it's already here. We are to live lives marked by joy and delight, to demand justice and mercy from our leaders, and to model lives of contentment and fulfillment. In short, to live as if God's tomorrow is today.

We live this way not because we believe that by doing so we can somehow bring God's coming future faster. We do so because, by living like we believe the end of the Big Story is that God has eternal plans for the created world, we are being faithful. We get better—through fits, starts, and stumbles—at loving God, loving God's world, and loving our neighbors. We get better at following Jesus.

As we cling to the "already" while firmly ensconced in the "not yet" of God's kingdom, we have this treasure in jars of clay. The bone-deep trust and overwhelming hope that our efforts to love God, God's world, and our neighbors by addressing climate change matter. In the economy of God's certain good future, every act of faithful resistance is a down payment on a future when resistance will no longer be necessary. Every prophetic call to action is an investment in a future where prophets will rest their voices at last, when God will be all in all, and the knowledge of God will cover the earth as the waters cover the sea.

This future is coming. Of this we can be sure. And this hope gives us the strength we need to live now as though it is already here.

~

This is the story that shapes our lives and orders our steps. This is the Big Story of God's deep and abiding love for all of his creation, our special role of responsibility in the midst of it, and his intentions—through Jesus—to bring all of it back to himself once more. We made many stops on our journey through the story, and we could have made many more. We could have stopped in Hosea 4, Joel 1, or any number of other minor prophets that remind Israel to do right by both the land and people, noting that oppression of one means suffering for both. We could have discussed the teaching in Romans 8 that all of creation is groaning under the weight of sin and death and waits on tiptoe for its human caretakers to set it free. We could have explored Ephesians 1, so like Colossians in its teaching that the mystery of God's will is to gather up all things in heaven and on earth through the good pleasure of Jesus.

Scripture shouts the Big Story of God's love for all of creation, our responsibility toward it, and the good plans God has for it. Somehow, this full-throated version of the Big Story has become narrowed in the Western church's retelling and has become entangled with all the other stories we explored in chapter two. Yet, far from being a small trickling stream tucked away in a few dusty pages of Genesis, the theological imperative to love and serve God's good creation courses from Genesis to Revelation, and cuts right through the heart of the gospel itself.

So how can we begin again to tell the fullness of this Big Story to a world desperate for good news?

4

CLIMATE ACTION IS GOOD NEWS

WHEN I WAS IN HIGH SCHOOL, I interned for a summer at my church. One day, a young guy about my age came to the building looking to talk to someone. The adult staff member who had met him at the door came back to the office where I was working and told me to come with her. She led me to a table where Ryan (not his real name) was sitting, eyes downcast. "He wants to talk," the staff member told me with knowing eyes. The subtext was clear: Ryan might not know Jesus, and it was my job to facilitate an introduction.

Ryan told me he was going through some hard stuff. I tried to listen. I told him Jesus loves him and wants to know his hurts too. I felt profoundly awkward, but Ryan seemed receptive. We exchanged numbers. He came to church a couple of times that summer, but one time we took Communion and all the talk about eating and drinking Jesus spooked him. We lost touch.

I still think about him.

Do You Know My Friend Jesus?

Evangelism loomed large for me growing up. Whether from the pulpit on Sunday morning, my youth pastor on Sunday evenings, or any number of teachers at my Christian school, the message was clear: evangelism is one of the most important things we can do to express our faith in Jesus and our gratitude for the grace we've received—to which I think most reasonable Christians would say, "Amen!"

I had friends who were amazing at it. I, on the other hand, always seemed to shrink from opportunities to share the reason for the hope that I had. I squirmed with discomfort when my youth group leader would ask us if we had found opportunities to share the gospel that week, averting my eyes and trying my best to blend in with my carpeted-gym-floor surroundings. I wrestled mightily within myself whenever I spent any meaningful amount of time with strangers, trying to work up the courage to share Jesus with them and then excoriating myself as they walked away, convinced I had disappointed Jesus and failed as a Christian. For much of my adolescent years, I felt enormous pressure to evangelize and shame when I would inevitably fail to wave the Jesus banner high in every aspect of my life.

And when I did find the courage to share Jesus with those around me, as with Ryan, the effects were decidedly mixed. It's possible that the Holy Spirit has worked in Ryan's heart since we last met and that my inelegant attempts at evangelism nevertheless planted a necessary and precious seed. It seems just as likely, though, that Ryan's brief experiment with Jesus that summer confirmed his worst suspicions about those crazy, body-and-blood-eating Christians and that my halting efforts to woo him into the story of God felt more like a cynical pitch than a winsome invitation.

When I went abroad with my youth group the same summer I met Ryan, the results weren't much better. We traveled to Tecate, Mexico, on a short-term mission trip, where I experienced profound spiritual growth and my global horizons were expanded. But I'm not at all sure our efforts brought anybody to Jesus. For one thing, demographically speaking, nearly everybody we encountered there was almost certainly already Pentecostal or Catholic. For another, our methods in retrospect feel pretty tone deaf to the realities we encountered on the ground: sixteen-year-olds slapping together prefabricated sheds to house families while skilled local laborers searched in vain for work; White teenagers snapping picture after picture of themselves with Brown kids in search of the perfect Facebook profile picture. While short-term mission trips certainly have their place and while I hope this one did some good, I can't help but admit to myself that when it came to the out-of-work laborer watching us encroach on his job market or the mother watching us exploit her child for social media likes, it's entirely possible that we did more harm to the Jesus movement than good.

I don't think I'm the only one with these kinds of stories. And it has me wondering: Is this all there is to this whole evangelism thing? If my only choices are either to squirm in self-hatred every time that I fail to insert Jesus into a casual encounter or to be laughed off Daytona Beach, count me out. Put another way: How might proclaiming the good news of the Big Story of God's world-redeeming work change the way we think about evangelism in the first place, especially in a warming world?

Euangelion and the Kingdom of God

The English word *evangelism* comes from the Greek word *euangelion*, meaning "glad tiding" or "good message." Though it comes to be used later in the New Testament to describe the saving work of Christ on the cross, *euangelion* shows up first in the Gospels themselves.

The Synoptic Gospels (Matthew, Mark, and Luke) all frame Jesus' earthly ministry as a proclamation of "good news." Each of them either describes Jesus as proclaiming the "good news" (Matthew 4:23) or puts the words "good news" in Jesus' own mouth (Mark 1:15; Luke 4:18). And until Jesus' death, resurrection, and ascension, this "good news" is understood not as the good news about Jesus but rather as the good news about the kingdom of God.

Jesus spends more time during his earthly ministry talking about the kingdom of God than about anything else. He tells no fewer than seven parables describing it. The Sermon on the Mount is Jesus' teaching on the ethical shape of the kingdom of God—a reality in which evil is returned for good, where worry is swapped for abiding trust, and where those who are poor and who suffer for the sake of what's right hold all the cards. It is, in short, God's ultimate plan for creation—that coming good future that John holds out to us in Revelation 21.

Jesus is obsessed with the kingdom of God (or "kingdom of heaven," since both are used interchangeably in the Gospels), because the kingdom of God is the bedrock of the Big Story of God's saving work in the world. God is inaugurating a new reality, through Jesus, which is good news for the entire creation. The kingdom of God is ushering in a world-altering, reality-bending, seismic shift in the way the world works, a radical subversion of the status quo where the first shall be last and the last shall be first.

Good News for Whom?

If the kingdom of God promises to upend the current status quo, the question must then be asked: What is the current status quo of a world gripped by sin? It is a reality where wealth and power are accumulated by a privileged few through the exploitation and oppression of the many. It is a reality where any piece of creation deemed to be inferior or subservient to humans and human needs is desacralized, instrumentalized, and sacrificed in the name of human progress, comfort, and pleasure. Cruelty is commonplace, selfishness is deified, power is a zero-sum game, and mammon is God.

In the twenty-first century, the runaway warming of God's creation is a potent, physical, and global manifestation of the sum total of all of this brokenness. And like so many other injustices, it is harming people at the bottom of society more than those at the top. Why? Because climate change is rapidly changing the basic operating system of the world, making it fundamentally more dangerous and unpredictable. It takes access to resources and durable institutions to adapt to a world that's shifting beneath our feet—access that those who are poor and oppressed already have precious little of (more on this in the next chapter).

Wouldn't it make sense, then, that the *euangelion* of a world flipped right side up would be particularly good news for those pressed to the bottom by climate change's cruel and unequal calculus? Jesus seems to think so.

In Luke 4, Jesus inaugurates his earthly ministry. He's returned to his hometown synagogue, and he has the honor of offering the public reading of the Torah. Jesus reads from Isaiah 61:

> The Spirit of the Lord is on me,
>> because he has anointed me
>> to proclaim good news to the poor.
> He has sent me to proclaim freedom for the prisoners
>> and recovery of sight for the blind,
> to set the oppressed free,
>> to proclaim the year of the Lord's favor. (Luke 4:18-19)

Then, sitting down with every breath held and every eye fixed on him, Jesus says simply, "Today this scripture is fulfilled in your hearing" (Luke 4:20). God's promise spoken through Isaiah centuries ago that he would privilege those at the bottom of society by upending the status quo, says Jesus, had been kept. Jesus himself would be this good news, the implications of which would follow wherever Jesus went. The table-turning, power-structure-subverting kingdom would break in with every mercy shown to the powerless and every rebuke of the powerful. Jesus' ministry had its mission statement, and the good news that he had come to proclaim was coming into focus: dignity and relief from suffering for those society had thrown away.

Jesus could have chosen any number of ways to inaugurate his ministry, any number of ways to announce what he had come to do. Yet, he chooses to quote a prophet, to center the poor and the oppressed, and to make perfectly clear that if the results of his ministry on earth aren't good news for those at the bottom, then he will have failed. The barometer for success, says Jesus, is not butts in seats, number of sermon downloads, or cash in the collection plates. It's release for the captive, freedom for the oppressed, hope for the hopeless.

In other words, if your ministry in the name of Jesus doesn't result in actions that are good news for the poor, you're doing it wrong.

In a warming world, this means that if our proclamation of the gospel doesn't result in actions that are good news for those suffering under the burden of an unequally changing climate—good news for those choking on pollution, losing precious farmland to rising seas, languishing under oppressive heatwaves, escaping deadly disaster and wasted homes only to run headlong into the arms of traffickers—then we, too, are doing it wrong.

Who Will Be a Witness?

We all know the famous instructions of Matthew's Great Commission: "Therefore go and make disciples of all nations, baptizing them in the name of the Father and of the Son and of the Holy Spirit" (Matthew 28:19). It's notable, though, that Mark, Luke, and Acts all offer their own marching orders too for how Jesus' disciples are to proclaim the good news. Like a diamond refracting light in myriad directions, we can get a more robust understanding of our calling to share the good news in a warming world by viewing the question from all angles.

In Luke and Acts, Jesus does not immediately send his disciples on the offensive as he does in Matthew but instead tells them to wait where they are. Only once they have received the Holy Spirit from the Father will they then be Jesus' witnesses in Jerusalem, Judea, Samaria, and the ends of the earth (Luke 24:46-49; Acts 1:8). Peter, Paul, and the rest of the apostles will go on to describe themselves and other Jesus followers as witnesses ten separate times in the book of Acts.

The word *witness* here comes from the Greek word *martys*, from which we also get the English word *martyr*. Through their words, deeds, and even their silence in the face of death, the martyrs of the early church were witnesses to the transforming power of the Holy Spirit in the world. The primary role of a witness, after all, is to watch and tell about the actions of another. Martyrs, apostles, and evangelists are not the evangelistic actors. According to Luke and Acts, the Holy Spirit is.

As for Mark, the "[*euangelion*] of God" (Mark 1:14) that Jesus began proclaiming through Galilee at the beginning of his ministry is a throughline that follows Jesus all the way to the very end. When Jesus commissions his disciples, he recovers the language of Mark 1 by telling them to "go into all the world and proclaim the [*euangelion*] to all creation" (Mark 16:15).

According to Mark, this good news is meant not only for those bent low by the oppressive and exploitative structures of the present evil age but for the entire groaning creation. Another faithful translation could render it as "proclaim the good news to every created thing in the cosmos." I wonder if Paul had Mark 16:15 in mind when he penned his Colossian Christ hymn.

Both Mark and Paul seem to be in sync with Saint Francis of Assisi, at least, the Catholic friar who would be born over a millennium later. Francis seemed to internalize the call to proclaim the good news to the whole creation more profoundly than most. He was famous for preaching to the birds that would roost in the trees along his walking path. Francis is even said to have saved a local village from being terrorized by a wolf simply by befriending it. And he spoke of creation as family, referring to Brother Sun and Sister Moon, Brother Fire and Sister Water in his Canticle of the Sun. St. Francis took Mark at his word and made a life of proclaiming the good news to all creation.

If Matthew 28 were the only account in which Jesus offers instructions to his disciples before his ascension, then the laser-focused emphasis on word evangelism that I and so many others encountered in the church growing up would make perfect sense. Yet, Mark, Luke, and Acts offer us additional insights to round out a more robust vision of what it might mean to, like Jesus, preach the good news of the kingdom of God.

Taken together with Jesus' mission statement in Luke 4, the full witness of the Gospels is that the good news of Jesus is meant for both all people (Matthew 28:19) and all of creation (Mark 16:15), that it is particularly good news for those who are at the bottom of the world's current hierarchy like the poor, the disabled, women, and people of color (Luke 4:18-19), and that it is spread by the Holy Spirit working through us (Luke 24:44-48 and Acts).

Good News for a Warming World

In a world where all people and all creation are being harmed by an out-of-balance climate, and where already vulnerable people are being hit hardest and fastest, what will it mean for us to bring this kind of good news into all creation? For starters, it means that if the gospel we share is meant to be good news for those at the margins of society, then we better start asking ourselves: Does the good news we have to offer actually result in things that are good for those who suffer? I don't mean "good things" like a pious platitude or a casserole every now and then. I mean paid bills, food in stomachs, and water that's safe to drink. Real, concrete *euangelion* for precisely the people whom Jesus himself chose to put at the center of his story of cosmic redemption.

Jesus had a habit of giving people a taste of the kingdom here and now. He healed people immediately. If someone needed to turn away from something that would prevent them from experiencing the fullness of the kingdom, he told them to repent right now, not later. To the rich man, he said to sell all his possessions now and follow after him (Matthew 19:16-30). Jesus was impatient for the kingdom. He wanted everyone to experience it as soon as possible—even if only provisionally before it arrives in all its fullness.

Jesus' ministry—his evangelism—was marked by belief in the radically present kingdom. He never told the disabled to wait for full healing. He told them to pick up their mat. He never told those who were hungry that he would pray for them. He multiplied loaves and fishes until everyone was satisfied. He never told those suffering from the consequences of systemic oppression, "Make peace with your suffering now because your reward is waiting for you someday in heaven." He healed lepers and bleeding women so that they could present their cleanliness to the priests and society would stop excluding them immediately.

In short, his actions in the world in the name of the gospel were good news for those who suffered. His evangelism, through his miracles and the rest of his ministry, represented the real in-breaking of the kingdom of God in the middle of history. It transformed lives. It upended entrenched systems of abuse and oppression. It challenged power—not in some far-off, someday reality but in this reality, now.

Does our evangelism in a warming world do the same? Does our evangelism lead to real, concrete good news for those enduring food insecurity due to changing weather patterns? Does our evangelism comfort Black mothers who have to watch their babies struggle for breath as they suffer through another fossil fuel pollution-induced asthma attack? Does it stand in solidarity with them as they fight for their child's right to breathe free?

Does our evangelism lead us to support Midwestern farmers on the brink of bankruptcy due to flooding that just won't quit? What about the girl in Appalachia who has to miss dance class in order to drive to Charleston for her next round of chemo? Does our evangelism directly result in their liberation, here and now?

If the first mark of evangelism in a warming world is that it follows the example of Jesus by acting in ways that are good news for those bent low by the impacts of a changing climate, then the second mark is that it de-centers ourselves and re-centers the Holy Spirit. It can be tempting to believe that the work of spreading the good news of the kingdom to all creation is all up to us. After all, Christ has no body on earth but ours, said Saint Teresa of Ávila. We must be his hands, his feet, and his voice in a warming world desperate for news that is good. And indeed we must, but not because the kingdom depends on our efforts. This, after all, is not what Jesus told his disciples in Luke and Acts. The very first step in Jesus' "Field Guide for Effective Evangelism" in these accounts is to wait. His disciples were instructed to wait for the Holy Spirit, and only after they had received it could they then become his witnesses. The Holy Spirit is always the primary evangelist. We are only ever her assistants.

When we are recast in our proper role as supporting cast rather than leading players, our evangelism loses the anxiety that so plagued my evangelistic experience growing up. The eternal destiny of all people and all creation does not, after all, rest on our shoulders alone. This anxiety can then be replaced by curiosity as we strive to understand what good news looks like in particular places and in particular contexts. Rather than assuming we know what is best for those who are suffering the impacts of a changing climate, we can listen and take our cues from them instead. In partnership and authentic relationship, we can join together to enact the good news of the kingdom of God.

There is a story in Barbara Kingsolver's *Poisonwood Bible* that I've never been able to forget. The novel is a tale about a Baptist missionary from the United States who, filled with evangelistic zeal, moves his young family across the ocean to the Democratic Republic of Congo. It is the 1950s, and revolution is brewing. The region has lived for decades under Belgian rule, first as the private nation of Belgian King Leopold II and later as a colony under the authority of the Belgian parliament. The natives have endured some of the most brutal conditions of any colonized people anywhere in the world.

The missionary, however, is blind both to this history and to its implications for the present. He is focused on one thing, and one thing only: getting as many Congolese people baptized as possible. He preaches with fire and conviction, and week after week, he ends every single one of his sermons with an invitation to join him at the river and to be baptized. And week after week, he stands alone on the riverbank as the local villagers file out of the church tent and make their way back to their homes. The missionary's failure to baptize a single person eventually causes him to have a nervous breakdown and to journey into the heart of the jungle alone, never to be seen again.

It is only later that his family discovers the reason for his evangelistic failure. It was common knowledge among locals that the river running by the village was infested with crocodiles. No one ever entered the river. The missionary's metaphors to "give up one's life in the waters" and to "die to Christ" were taken literally by the villagers. They were convinced he was a murderer, intent on killing them and their children.

The crocodiles in the river, the brutal history of Western imperialism, the specter of revolution that would touch the life of each and every person in the village—the missionary was blind to all of it. He had evangelistic tunnel vision. His only conceivable metric for "success" was the number of souls he could manage to plunge beneath the crocodile-infested waters. And because he had forgotten his place subordinate to the Holy Spirit, the pressure of his perceived failure broke him.

If evangelism in a warming world is marked by acts that are good for those who suffer and is grounded in the primary action of the Holy Spirit, then its third mark is that it is proclaimed to every corner of God's broken and scarred creation. When the task of proclaiming the good news in ways that result in good things is extended to every corner of creation, then suddenly the work of evangelism becomes expansive. We find ourselves invited less into discrete acts of evangelism and into a good news–shaped way of life. Our habits of consumption; our relationships with neighbors and friends; our acts of citizenship; the ways we dress, eat, and recreate; and how we invest our time, money, and talents all become opportunities to be a witness to the world-transforming reality of the kingdom of God.

In a world wracked by the devastation of rising seas, more extreme weather, and a more unpredictable and dangerous future, this means that practical and meaningful steps to address the climate crisis are, in fact, acts of evangelism. If the good news of the radically near kingdom of God is meant to be proclaimed to the whole of creation, then plunging our hands into the dark, rich soil of our gardens every spring and volunteering to remove invasive species from a local forest are acts of defiance against the forces of death and an enactment of the hope that we have for God's coming new creation.

If the gospel of justice, wholeness, and life is meant to have particular force for those whose lives are marked most heavily by injustice, disintegration, and death made even heavier by pollution and a changing climate, then investing in local food sovereignty projects becomes a witness to the hope that we have in the coming future where Christ will lay his banquet table, and all will be welcomed to it. Joining the work of environmental justice groups to fight for the dignity and self-determination of local communities becomes a foretaste of the day when all will flourish in health and wholeness in God's coming good future. Advocating for public policies that rapidly draw down green-house gas emissions, invest in communities harmed most by climate pollution, create family-sustaining jobs with livable wages, and pave the way toward a healthy economy powered by clean energy becomes a blow against the death-dealing powers and principalities of the current age that seek to entrench the status quo and forestall the inevitable consummation of the kingdom of God on earth.

All these actions can, of course, be undertaken by any number of people for any number of reasons. There is nothing inherently evangelistic about gardening or advocating for public policy. But when Christians do these things, with the hope of God's coming kingdom in our hearts and the name of Jesus on our lips, they become proclamations of good news for the whole creation. When we engage as Christians in public actions whose results are good news for the poor, for the oppressed, and for all of creation, then we bear witness to a watching world to the reason for the hope that we have in Christ. When we take private steps as Christians to better the condition of the natural world around us, we announce to a world weary with death that its present groaning will not be permanent and that we stand on tiptoe with it in expectation of God's coming redemption.

Along the way, we must never be afraid to share the name of Jesus with any and all we may meet. After all, we will merely be following Jesus' instructions. Yet, when we do share the name of Jesus—whether it is emblazoned on our shirts as we march for climate justice, woven into our arguments in the opinion pages of our local newspaper, or whispered softly to our tomato plants as we gratefully receive their ruby gifts—we will do so not out of an anxious obligation to hit a conversion quota or because we alone stand between a stranger and her eternal damnation. We will do so freely, joyfully. We will do it out of the freedom of knowing that we are called not to keep a running tally of souls saved but to bear faithful witness to the goodness of God's coming kingdom and to let the Holy Spirit do the rest. We will do it because climate action is indeed very good news.

5

BEING PRO-LIFE IN
THE AGE OF CLIMATE CHAOS

THE SUN BEATS DOWN as we pick our way through the lush under-growth. I see her up ahead of us, hunched slightly with age and leaning on her hoe. We trudge closer, sweat pouring down our backs, until we enter her garden.

But to call it a garden isn't quite right. It's a veritable farm plot, wild in all its rainy season decadence yet expertly cultivated. Margaret considers each of our White faces in turn, her welcoming smile never faltering.

We've come to Margaret's patch of land for a story—her story. Stories are the reason, in fact, that our group of Americans and Canadians has been crisscrossing the Kenyan landscape for close to a week. We've traveled from verdant hillsides to jam-packed slums and from the blaring cacophony of downtown Nairobi to the peaceful plains of Nakuru to receive firsthand knowledge of how changing weather patterns and a warming world are already pressing on people on the other side of the world and to bring those stories back home with us. In return for receiving these sacred gifts, we have been making a promise to those who offer them to share these stories with our friends and families—to present them to our churches, to our classmates, to anyone that might listen—in order to bear witness to the daily realities of our brothers and sisters in Christ in Kenya and across the world. It's our own, small way of showing them that we believe them when they tell us that climate change is killing them and to do what we can to stir our Christian communities back home to engagement and action.

Margaret waits placidly until all our group has joined the semicircle surrounding her and have caught their breath. Then she opens her mouth and, with the help of an interpreter, shares her gift with us.

She's lived on this land for a very long time, she tells us. She's grown up with the knowledge of the land in her very being. As a child, she had watched the adults of her village enact the subtle give and take of their marriage to the land. They would prepare the soil, plant the seeds, and pray. Their faith would be rewarded as the heavy spring rains soaked the earth year after year. Margaret learned the lessons of her way of life less through explicit instruction than through regular practice. She would accompany the adults into the fields, watch quietly as they went about their work, and listen intently to their conversation. Slowly, stumbling and tripping at first and then more gracefully, she picked up the steps of the dance.

She grew on that land, alongside the maize and the papaya trees. As an adult, she continued planting and harvesting, enacting the rhythms of life with her own children underfoot. But something was amiss. The other adults had noticed it too. The rains, so dependable in her childhood that their arrival could be planned almost to the day, were now harder to predict. Somehow, the intimate dance of land, rain, and farmer—for so long in lockstep—had grown out of sync. Years passed, and, try as she might to find the rhythm once more, the dancers grew farther apart.

Now, Margaret tells us, her dance partner has become a near stranger. The rains, she says, are unrecognizable and have become almost impossible to predict. Farmers in her village, once so attuned to the life of the earth and the rains, are now left to make their best guesses and to pray to the heavens. Unlike their parents and grandparents before them, their prayers are being answered less and less.

Some years, Margaret tells us, a farmer might plant and the rains not come for an entire month. Once the rains finally do come, the seeds will have all too often given up their waiting and will be long dead beneath their burial shroud of soil. Chastened by the hardship of missing an entire growing season, a farmer may decide to wait to plant the following year, only to watch the rains come early and wash away another opportunity to feed her family. Some years, says Margaret, the rains will come all at once and carry their seeds down the hillside in torrents. Other years, the rains might decide not to come at all.

This distortion of the dance takes its toll. Counting her own children, grandchildren, and neighboring children she has taken in, Margaret is responsible for feeding thirteen mouths. When the music was predictable and the dance steps ones she knew, she was really good at it. It was only when the tempo changed and the steps got messy that she began to stumble.

Not far from Margaret's village is a market. Growers and vendors from all over the region make their way to the row of stalls on market days to hawk their wares to the neighboring villagers. The dazzling yellows and golds from the sun-ripened mangoes and maize blend beautifully with the deep purples and ruby reds of kitenge fabrics dotting the stalls.

Margaret had a stall here once. After she had fed all the mouths she had promised to feed, it was here among the bustling shoppers and shouting sellers that the vibrant colors from her own excess produce joined the exuberant kaleidoscope of abundance. The extra money she made here would pay school fees, buy new shoes, or purchase treats for herself and her family.

But no more. Once the rains introduced their foreign steps into the sacred dance of survival, Margaret wasn't able to keep up. Facing mounting food shortages and hungrier bellies at home, Margaret was forced to supplement her family's income by finding a new, harsher dance partner. As a day laborer, Margaret goes where there is need of arms to lift concrete bags or legs to pedal messenger bikes. On payday, she makes her way back to the market, a path her feet have beat countless times. Yet, this time she does not take her place behind a stall, the precious fruit borne of her intimate partnership with the land spread before her. Instead, she walks the paths between the stalls, her hard-earned money held tightly in her hand. She repeats her total earnings in her head like a mantra. She ceaselessly calculates the prices she sees against the meager wad clenched in her fist, praying that this time she will have enough.

Day after day, Margaret's body enacts the monotonous rituals of hard labor. Yet, her mind and her spirit are far off. They often fly from her work, over the dust-covered roads and back to her sun-drenched field, still wet from the day's heavy rains. She walks amid her garden, lightly touching maize leaves as she goes. She revels in the wild abundance, gifts from the land that she loves. And for a moment she can hear the music again. The dance comes back to her, and she is free.

～

When I was in middle school, my Bible teacher gave our class the assignment of presenting to the class a Christian perspective on a contemporary political issue. We were free to choose our subject of study, but she wisely set a limit of two students per topic so as not to endure a parade of presentations on the same theme. I was one of the lucky ones. After a small scrum at the sign-up sheet, my name could be found under the most coveted issue of them all: abortion.

I embraced the project with relish. I scoured the internet, looking for the most damning statistics and the most gruesome anecdotes I could find. It was almost too easy. I had been primed for this moment.

For my entire life to that point, abortion had dominated my political formation. It was the primary political concern of my Christian community above all others. For much of my growing-up years, it was almost larger than life for me. It was somehow both the cause and the effect of the moral degeneration of America, the root of my generation's slide into moral relativism, and the reason our society was hurtling toward a godless nihilism from which only the faithful could pull it back.

And so, I threw myself into the project with gusto. My arguments were ironclad. My evidence was irrefutable, and my rhetoric soaring.

And most importantly, my cause was just. I got an A on the project and a standing ovation (okay, maybe just the A).

We touched briefly in chapter two on why abortion dominated so much of my cohort's political formation growing up in 1990s and early 2000s evangelical culture. And I have deep respect for those who advocate for an end to abortion—many of them my own friends and family. But I can't help lamenting that for so many a Christian political commitment to life begins and ends with advocating for the legal abolition of abortion. Of course, many pro-life Christians couple their anti-abortion advocacy with acts of private service like volunteering at crisis pregnancy centers and even opening their homes to children in need. Yet, when it comes to political action in the public square, pro-life Christians are often singularly focused.

There is nothing wrong, of course, with Christians expressing their passion for life in this way. What is problematic is how, for so many, this is the extent of their Christian civic witness. Yet, there are myriad policy issues that affect people's ability to experience abundant life—access to safe and affordable housing, criminal justice reform, food sovereignty, access to affordable healthcare, police reform, racial justice, and so many more. And, of course, underlying all of these issues is access to a healthy environment and a stable climate.

If I have heard one message more than any other in my years organizing, speaking, teaching, and advocating with Christians for climate action, it's that climate change has nothing to do with Christians or Christian discipleship. However, this can only be true if climate change remains a purely theoretical abstraction. When we meet real flesh-and-blood people whose lives are being threatened by it—people like Larry and Margaret—climate change becomes much harder for the Christian to set aside.

The truth is, the Big Story of God's saving work in the world, and our calling to proclaim its good news to all creation, has everything to do with climate change. Our faith not only permits but requires us to clean up toxic waste sites, protect our neighbors from harmful pollution, and safeguard God's human and nonhuman creation from the ravages of a changing climate.

So, what might it look like for followers of Jesus to expand the scope of what it means to be pro-life in a warming world? What if the reputation of Christians in the public square included more than merely their anti-abortion fervor but also a holistic concern for all life at all of its precious stages?

Climate Change Is Killing Certain People More Than Others

I've often wondered about the peculiarity of climate denial. It's not a universal response to the overwhelming evidence for human-caused climate change. In almost every country on earth, the scientific findings of anthropogenic global warming are as accepted as the laws of gravity or germ theory.

As a civil society observer at the 2015 COP21 UN Climate Change Conference in Paris (where the Paris Agreement would be reached), one of my most searing takeaways was the sheer universal acceptance of climate science among the people of the world. Even Saudi Arabia, whose entire national existence depends upon the ongoing sale of crude oil, was at the table. Granted, they were doing all they could to water down the agreement language and slow walk diplomatic progress. Still, they weren't calling climate change a hoax. It's only in a handful of wealthy, industrialized countries where doubt and denial of the scientific facts persist. Perhaps nowhere does this privilege of skepticism persist more strongly than in the United States.

There are lots of cultural, economic, and political reasons for this that we touched on in chapter two and that could (and already do) fill another book. But there's one reason that often gets overlooked: geography. The North American continent is situated almost entirely between the North Pole to the north and the equator to the south. Sandwiched in the middle of the continent is the contiguous United States, cozied up roughly between the twenty-fifth parallel at the northern edge of Florida and the forty-ninth parallel along the Canadian border.

If you're like me, you probably haven't thought about latitudes and longitudes since the fourth grade. But when it comes to climate change, where the latitude/longitude lottery has seen fit to plunk you and your loved ones down on planet Earth can have life or death consequences. This is because climate change presses on certain parts of the world more than others, and it does some of its hardest pushing at the poles and at the equator. At the poles, climate change manifests most powerfully in warmer temperatures. The Arctic, in particular, is warming faster than any other part of the world, and about twice as quickly as the global average.[1] In June 2020, the Arctic Circle recorded its first-ever 100° Fahrenheit day. The following spring, hundreds of wildfires burned across Siberia and in June 2021, a European Union satellite recorded ground temperatures in Siberia's Sakha Republic in excess of 118° Fahrenheit.[2]

This has terrible consequences for the people who live in these extreme north regions, not to mention the global risks it poses by way of positive feedback loops as vast swaths of baking permafrost thaw and release massive amounts of methane into the atmosphere, thereby driving further warming. In a truly horrifying twist, dormant diseases locked away in permafrost appear to be reawakening and infecting humans too, like the anthrax outbreak in northern Russia in 2016 that was linked to a seventy-five-year-old caribou carcass that had been thawed by the extreme temperatures.[3]

The truth is, though, a relatively small proportion of the world's population lives in the Arctic Circle. The equator is a different story. Like the poles, the equator experiences some of the most extreme climate impacts—namely, wet tropical regions like Indonesia and the Philippines are getting wetter while arid regions like sub-Saharan Africa are getting drier. Unlike the poles, 79 percent of the world's population lives in countries closer to the equator than Japan, or roughly within the northern and southern thirtieth parallels. This includes the Indian subcontinent, Southeast Asia, almost the entirety of Africa, Mexico, and Central America, most of South America, and much of northern

Australia. And though these regions hold most of the people on earth, they hold only 31 percent of the world's wealth.[4]

This means that the majority of the world's people who are feeling the strongest impacts of climate change are also those with the fewest resources to be able to adapt. And, since economic development and wealth accumulation in the nineteenth and twentieth centuries were directly tied to how many fossil fuels your economy consumed, it also means that, historically, these populations have contributed least to the problem of climate change. Indeed, of the top-twelve highest historic CO_2 emitters from 1850 to 2011, only one (India) sits between the thirtieth parallels.[5]

Of course, this is not to say that climate change doesn't press on those of us outside these boundaries too, particularly in more recent years. Heat waves have killed hundreds in Europe and the Pacific Northwest of the United States. Flooding has ravaged Germany, China, and much of the United States. Wildfires in Greece, Algeria, and the American West have devoured millions of acres and crumbled countless homes and livelihoods to ash. Stronger and wetter hurricanes, driven by warmer air and warmer oceans, have crippled not only southern US cities but also northern cities like New York. Rising coastal waters aren't just a problem for Miami; they also threaten the Naval Station Norfolk in Norfolk, Virginia—the largest naval base in the world. The US Department of Defense is already implementing adaptation efforts at the base, such as dune restoration and floodwall construction in order to, as Defense Secretary Lloyd Austin put it in 2021, "mitigate against this driver of insecurity."[6]

However, thanks to that lion's share of global GDP, many populations outside the thirtieth parallels are better equipped to adapt to many of these emerging realities. The US Navy has all the money it needs to retrofit Naval Station Norfolk. Not all, but disproportionately more residents of Western and Northern Europe have the financial means to purchase air conditioners to beat the heat than do their counterparts in Ghana or India. Farmers in the United States have access to drought-resistant crops, university extension programs, and government-backed crop insurance of which their peers in sub-Saharan Africa can only dream. The Netherlands has been adapting to marauding waves for millennia. Their relative wealth, accumulated in part through colonial exploitation and the lucrative Transatlantic Slave Trade, has enabled them to construct elaborate networks of seawalls that are true engineering wonders to behold. In contrast, similarly low-lying Bangladesh had their wealth flow in the opposite direction across colonialism's cruel ledger and is now left to cede more and more of its arable land to the sea's insatiable lapping each year. It is no accident, after all, that these regions closest to the equator have so much of the world's population yet so little of the world's GDP. They were also the most heavily and disproportionately colonized by those nations outside the thirtieth parallels—namely, Western Europe and the United States. In the great formula of climate injustice, colonial theft plus geographic vulnerability to climate impacts equals the greatest climate suffering.

Here in the United States, the same pattern of climate change's unequal footprint holds true. Among our post-industrialized peers, no country has a higher level of income inequality than the United States. "In 2018, the top 10 percent of U.S. households held 70 percent of total household wealth, up from 60 percent in 1989," according to the Federal Reserve.[7] Under this kind of extreme economic inequity, it necessarily follows that there will be an ever-shrinking minority of Americans who will always be able to escape the deadly impacts of climate change largely unscathed, and a growing percentage of Americans who cannot.

The early days of the Covid-19 pandemic are illustrative of how economic inequity plays out in the face of a crisis in real time. It was well documented that during the initial, devastating wave that hit New York City in the spring of 2020, many New Yorkers fled the city to try to escape the pathogen, but not all of them could afford to leave. As wealthy Manhattanites fled the carnage to second homes outside the city, those without the means to escape were left to fend for themselves. The data from that first wave tell us what we already intuitively know:

incidence, hospitalization rates, and mortality were highest among Black and Hispanic New Yorkers, as well as those living in neighborhoods with high poverty, those over seventy-five years of age, and those with underlying health conditions.[8] In other words, the virus most affected those without the luxury of simply picking up and moving away from the danger.

Overlaid on top of existing economic and racial inequality, Covid-19 took an existing problem and turned it into a full-blown crisis. This is what climate change does too. It is what the US military calls a "threat multiplier." It takes problems that already exist in any given society—economic inequality, health disparities, armed conflict, public health crises, poverty, hunger—and exacerbates them.

Perhaps the most famous recent example of climate change's power as a threat multiplier is the Syrian civil war. Syrian society had been building toward a breaking point for some time. The predominantly Sunni Muslim country had been ruled for forty years by the much smaller Alawite minority, which held onto power through an uneasy patchwork of alliances with other minority populations. Tensions simmered but remained largely in check until 2006–2010 saw the worst drought to strike the region in nine hundred years. Eight hundred thousand people lost their income, 85 percent of the country's livestock died, and food prices skyrocketed. Huge numbers of Syrian's flooded to cities to find work. The pre-existing conditions of poverty and religious and political tensions were supercharged by a massive spike in unemployment, mass migration, and a perceived indifference to it all from the al-Assad government. Climate change lit a match next to an existing tinderbox.[9]

While other examples may not be as dramatic, they are no less deadly. As fiercer and fiercer storms lash the US Gulf Coast, it will continue to be those with the resources to rebuild or relocate to higher and more expensive ground who will fare better than those who cannot. As extreme heat bakes the Midwest and boils urban centers, it is those with the means to buy air-conditioning units and homes in neighborhoods with tree cover and green space who will be able to escape the deadly temperatures. As long as torrential rains keep drenching the Great Plains, those with access to updated stormwater infrastructure and sump pumps will be okay. No matter how bad climate-driven droughts may get in the future, there will always be someone willing to load potable water into a truck and drive it until someone pays them for a gallon or two. I'll give you one guess who'll be able to afford it.

This is the fundamental injustice of climate change, and it is a direct challenge to those of us who consider ourselves pro-life. Although climate change is felt everywhere, geographic realities and socioeconomic inequities mean that its pressure is more deadly for some than it is for others. And it is disproportionately those who are least responsible for the impacts of climate change and who have the fewest resources to adapt who suffer the most.

The True Cost of Fossil Fuels

Of course, greenhouse gases are not emitted in a vacuum. When we dig up and burn coal, pump and process oil, and harvest and ignite natural gas, we are releasing much more than heat-trapping gases into our air and water. This means those of us who are concerned with protecting life have more than climate-driven extremes to worry about.

Heavy metals like lead, mercury, and selenium are all byproducts of coal production. Cancer-causing agents like benzene and other hydrocarbons are routinely released into the air by oil and gas producers through standard industry practices like venting excess methane into the air and flaring—the controlled burning of methane gas during the production process. Particulate matter measured at 2.5 microns or smaller ($PM_{2.5}$)—otherwise known as soot—is a ubiquitous byproduct of smokestacks and tailpipes alike and is strongly linked to increased risk of heart and lung disease, as well as a host of respiratory illnesses.[10] And this says nothing of the countless toxins that find their way into millions of hearts, minds, and lungs through other industrial processes— from PFAS (per- and polyfluoroalkyl substances) and other "forever chemicals" to the hexavalent chromium of Erin Brockovich fame. When we begin looking under the hood, it becomes immediately clear that our global industrial economy, driven by fossil-fuel energy, is being subsidized not only by trillions of government dollars every year[11] but also by the health and lives of millions of people around the world.

What's more, data and history tell us that this pollution does not impact all people equally. Just ask the residents of Warren County, North Carolina. More specifically, ask the poor, rural, Black residents of Warren County, North Carolina.

In September 1982 a familiar pattern was playing out in the small town of Afton, North Carolina. A hazardous-waste landfill had been sited in the poor, overwhelmingly Black enclave, and the first trucks were rolling in with their toxic payloads laced with polychlorinated biphenyls, otherwise known as PCBs. The local community had tried to communicate their concerns to state officials about the risks the landfill posed to their drinking water supply, but to no avail. Furious at the state's indifference and desperate to protect themselves and their families, they used the last resource they had at their disposal: their bodies. Over the course of six weeks, the residents of Afton and their allies blocked the trucks with marches and nonviolent protests. Over five hundred people would be arrested before it was all over—the first arrests in US history over the siting of a landfill.[12]

Communities of color had organized protests before to resist the threats of environmental harm. Perhaps most famously, Cesar Chavez and the United Farm Workers Movement in the 1960s struggled for protection from the harmful pesticides that farm workers were forced to apply day in and day out. But the Warren County protests captured national attention in a new way, sparking the imaginations of people who lived under similar environmental threats and helping to launch the modern environmental justice movement.

The environmental justice movement is a response to an eminently demonstrable reality: in the United States, poor communities—and especially Indigenous communities and other people of color—are disproportionately harmed by environmental pollution. There are many reasons for this.

Because of racist housing discrimination and zoning practices, Black and Brown communities are more likely than their White counterparts to be located near fossil fuel–burning power plants, toxic waste sites, heavy industry, and chemical processing plants.[13] These fenceline communities, so called due to their immediate proximity to these industries, experience disproportionately high rates of asthma, cancer, and other diseases linked to exposure to things like benzene, mercury, and soot.

These communities often suffer from a lack of public investment, leaving residents vulnerable to contamination from aging infrastructure and poor housing stock. In the case of Flint, Michigan—a majority-Black city with a median income almost half that of the national average and a poverty rate of 39 percent—a combination of failing infrastructure and undemocratic governing practices combined to poison an entire community. The fateful decision in 2014 to switch the city's water supply from the Huron River to the Flint River without adding corrosion control to the new water supply was made not by elected officials accountable to local voters but by an unelected emergency manager, who was legally appointed by Michigan's governor and whose authority superseded the elected city council. As a result, approximately 99,000 residents spent eighteen months drinking, bathing in, and cooking with lead-tainted water and couldn't even vote out the parties responsible.

During the construction of America's interstates in the 1950s and 1960s, an intentional policy was put in place to bisect Black neighborhoods whenever possible. Done in the coded language of "urban renewal" and "reclaiming blighted areas," the policy destroyed urban African American communities.[14] Those who remained in what was left of these neighborhoods were now not only denied the economic vibrancy of a functional community but also exposed to the dangerous soot and volatile organic compounds (VOCs) spewing from thousands of tailpipes zooming past their windows day and night.

Across the country, oil and gas pipelines are crossing sacred lands and poisoning the water supply of tribal nations with little or no legal consequences. Uranium mining on Navajo land from the 1940s to the 1980s, carried out in the name of patriotic resistance against the evil specter of communist aggression, poisoned generations. The treaty rights meant to protect Indigenous populations from this kind of exploitative endangerment have gone the way of so many other treaties made between Native people and the US government over the centuries—broken and trampled on in the name of economic and political expediency.

The list could go on—and indeed it does—but the point remains that while environmental pollution driven by the digging up and burning of fossil fuels harms all of us, it especially hurts those who are already vulnerable due to economic disenfranchisement and social marginalization. Those who already lack access to affordable medical care to treat pollution-induced conditions like asthma are unnecessarily harmed and end up suffering the long-term impacts of untreated disease. Those forced to live in urban areas with dense concentrations of roads, cities, and other buildings that absorb and retain heat experience precious little relief from tree cover or expensive air-conditioning. The elderly are disproportionately affected by temperature extremes because they often find it hardest to leave their homes to find cooling or warming shelters.

People with disabilities are not merely inconvenienced but threatened every time a power outage from extreme weather silences their life-saving medical devices. People living with chronic respiratory illnesses like COPD, whose lungs already struggle and strain for breath against the tightening bands of disease, are most at risk from pollution-laden air.

And it turns out young kids are some of the most at risk from the effects of fossil fuel pollution. There are several reasons, including this simple one: their lungs are smaller than an adult's. Smaller lungs require more frequent breaths and, therefore, are exposed to more of the toxins that may be present in the air around them like soot (PM2.5), ozone, and benzene.

We've already seen where PM2.5 comes from and what it can do to the human body. Ozone (also called smog) is produced when tailpipe and smokestack exhaust reacts with sunlight and heat. You know those hot, muggy ozone action days in the summer when everyone turns off their lawn mowers and rides the bus for free? It's to keep as much fossil fuel pollution out of the air as possible. That's because, when breathed in, ground-level ozone chemically attacks lung tissue. The American Lung Association equates breathing in ozone to "getting a sunburn on your lungs."[15] As for benzene and other carcinogens making their way into our lungs and bodies, oil and gas production are the main culprits. And again, it's young kids who are put disproportionately at risk. In the United States, more than three million children go to school within a half mile of active oil and gas wells, processors, or compressors.[16]

Fossil fuel pollution even affects the unborn. Early medical evidence is beginning to show that exposure to fossil fuel pollution while in the womb leads to disproportionately increased rates of low-birth-weight babies, heart and neural tube (i.e., brain, spine, and spinal cord) defects, and adverse health outcomes later in life.[17] Once many children are out of the womb, they fare little better. A terrifying preliminary study published in late 2020 shows that up to 90 percent of the world's children may be vulnerable to particle pollution (i.e., soot from fossil fuels), which is linked to increased rates of Alzheimer's and Parkinson's disease.[18] Like climate change, the health impacts of other forms of pollution that are emitted when we dig up, process, and burn fossil fuels are a challenge with which anyone who calls themselves pro-life must grapple.

According to the Centers for Disease Control and Prevention, there were 619,591 legal abortions reported in 2018. Though this number represents a 22 percent decrease since 2009, and though over 92 percent of these abortions were performed in the first trimester, it remains a heartbreaking number that many pro-life activists know by heart and that undoubtedly drove much of the activism that led to the recent reversal of *Roe v. Wade*.[19] Relatively few pro-life Christians (or anybody, for that matter) know a similarly heartbreaking number: close to two hundred thousand Americans also died in 2018—and die every single year—from air pollution.[20] As for the rest of the world, a study from *The Lancet* estimates that air pollution accounted for nine million premature deaths globally in 2019.[21] That means air pollution was responsible for one out of every six human deaths that year—three times more deaths than from AIDS, tuberculosis, and malaria combined and fifteen times more than from all wars and other forms of violence. Almost all of this pollution and death is caused by the digging up and burning of fossil fuels.

Pro-life advocacy has long been laser focused on ensuring all children have the opportunity to get that first gulp of precious air. But shouldn't it also matter to those of us defending life that for most children around the world, that first gulp—and every gulp thereafter—is toxic?

The Next Pandemic Is Closer Than You Think

Extreme weather and pollution are not the only climate-related threats to life in God's good creation. Climate change also spreads infectious disease. In particular, vector-borne diseases are having quite the moment in a rapidly warming world. Vector-borne diseases are diseases that result from an infection transmitted to humans by living organisms—usually blood-sucking insects—and cannot be transmitted human to human. Malaria, Zika, West Nile, dengue fever, and Lyme disease are some of the more famous vector-borne diseases. And all of them are getting worse because of climate change.

Lyme disease is quickly becoming an epidemic in the US Northeast and parts of lower Canada. Historically, the hard Canadian and New England frosts have been enough to kill any intrepid ticks exploring the wild frontiers of their habitat range and deter any others with half a mind to try. However, in 2016, the United States had fifteen more frost-free days on average than in 1895.[22] In Philadelphia alone, the date of the last frost of the 2019 season was roughly ten days earlier than it was in 1970.[23]

As temperatures climb and frost-free days increase, overall humidity also increases. This, it turns out, appears to be the key condition for ticks to flourish and thrive. A recent study found that in conditions of moderate (85 percent) or high (95 percent) relative humidity, ticks were much more likely to venture above the protective ground cover and seek out tasty morsels to bite than they were when relative humidity was lower (75 percent).[24] The species of tick that transmits Lyme disease (black-legged tick) makes its home in the eastern half of the United States (plus parts of California). Since climate-driven increases in relative humidity are less pronounced in the Southeast, Lyme disease is a particular threat to those living in the northeastern quadrant of US states. Indeed, in 2019, 85 percent of probable and confirmed US cases of Lyme disease occurred in New England, Midwestern, and Mid-Atlantic states.[25] If the world breaches the 2° Celsius threshold that nations are scrambling to avert, it's estimated that Lyme disease cases in the United States could increase by as much as 20 percent in the coming decades.[26]

Mosquito-borne diseases tell a similar story. As temperatures, humidity, and frost-free days increase in regions around the world, mosquito habitat expands as well. Malaria has long been endemic to much of sub-Saharan Africa. In Kenya, coastal and low-lying portions of the country have always been vulnerable, but higher-altitude Nairobi has historically been spared. As malaria creeps farther and farther into Kenya's capital city, high altitude is protecting it less and less in a warming world. Zika virus—long a scourge in tropical countries like Brazil—caused panic in the summer of 2016 when the first outbreak ever recorded in the United States was identified in Miami, Florida. West Nile virus made its first appearance in the United States in 1999.

In my home state of Michigan, we now have to worry about eastern equine encephalitis (EEE). EEE is a disease that originated in horses, is spread by mosquitos, and boasts a terrifyingly high 33 percent fatality rate in humans. Because the primary manifestations of climate change in Michigan (and the rest of the Midwest) are increased rainfall and higher temperatures, mosquitos are living longer and longer each season. In 2017, mosquitos had twenty-three more days to live in Detroit than they did in 1970.[27] Even before Covid-19 pandemic shutdowns were the norm, Michigan high schools were canceling late summer and early fall football games over EEE concerns.

Other diseases are also on the rise. Vector-borne diseases are a subset of a larger category of infectious diseases called zoonotic diseases. Zoonotic diseases jump freely between animals and humans. Rabies and Ebola are both examples of non-vector-borne zoonotic diseases—rabies being transmitted directly to humans from a nonhuman mammal host only and Ebola having jumped from animals to humans and now being transmitted human-to-human. Though it is as yet unproven, Covid-19 is believed to be in the zoonotic mold of Ebola—indirectly transmitted from an animal (potentially pangolins or bats)—and now spreading human-to-human.

Climate change will make these kinds of emerging zoonotic diseases more common too. As the impacts of climate change force unprecedented migrations of people across the globe, human pressures on animal habitat will only worsen. Rising seas and drought-induced agricultural collapse are already driving more and more people every year from their homes. At the same time, climate-driven increases in the frequency and severity of natural disasters like wildfires are gobbling up animal habitat. Climate change is shrinking the area of habitable land, and humans and animals are competing for the real estate that's left. As humans and animals come into closer and closer proximity to each other, zoonotic diseases will continue to jump from animals to humans with greater and greater frequency.

And as we saw with climate-driven weather impacts and pollution, disease is fundamentally unfair too. While diseases infect without discrimination, they kill with prejudice. Preexisting socioeconomic inequities and comorbidities will determine who is at greatest risk of severe disease and death from the zoonotic nightmares of the future. They already do for the diseases we face today. The unborn babies who are most at risk of Zika-induced microcephaly are found primarily in lower socioeconomic communities.[28] Frontline workers such as bus drivers and grocery store workers—disproportionately Black, Indigenous, and people of color—are put at greatest risk of exposure to Covid-19. Immunocompromised people are in greater danger of dying from Covid-19 should they contract it. And the vast swaths of impoverished people living at or near the equator still account for the overwhelming majority of malaria cases and deaths.

Being Pro-Life in a Warming World

Climate change is driving deadly weather events around the world, poisoning our air and water, and making us sicker by moving deadly diseases around the globe. More often than not, the weakest, poorest, and most vulnerable lives are those that are put at greatest risk. Being pro-life in the twenty-first century requires that we take these realities deadly serious because they are already deadly serious for so many people—our neighbors and ourselves. Yet, "pro-life" in the United States is not synonymous with "climate action." In many instances—particularly in the conservative Christian communities that tend to identify most readily as pro-life—they are ideological opposites.

It's high time for those of us who consider ourselves on the side of life, wholeness, and full flourishing to drastically expand our understanding of what it means to be pro-life. For too long, a relatively small group of powerful Christian leaders and their political allies have been allowed to hold hostage the moral imagination of millions of Christians. They've been allowed to circumscribe the definition of pro-life mostly to mean pro-birth, and they have blessed a very narrow set of public actions that can be pursued in its name: donate to and volunteer at crisis pregnancy centers, participate in the annual March for Life, and—most importantly of all—vote Republican.

With the Supreme Court's 2022 ruling overturning the constitutional right to abortion in *Dobbs v. Jackson Women's Health Organization,* this activism paid off. Anti-abortion activists achieved a generational victory. Yet, the confusion about the pro-life movement's post-*Roe* strategy since

Dobbs has been telling. The pro-life focus had been lasered in on *Roe* for so long that few seem to have asked what it means to be pro-life aside from *Roe*. For too long, the morally expansive question of what it means to faithfully honor the God-given gift of life in the public square has been reduced to a *Roe*-shaped binary. Perhaps now, like the proverbial dog who finally catches the car, the US church can ask anew what it might mean to advance a holistic ethic of life in a warming world.

As a political shorthand for abortion alone, the pro-life moniker is a remarkable bit of political branding. Yet, it fails to take into account God's expansive concern for all stages of human life. After all, Scripture is clear that God delights in all life—human and nonhuman—and is present at all stages of life. God "knit [us] together in [our] mother's womb" (Psalm 139:13), "will watch over [our] life" (Psalm 121:7), and finds "precious the death of his faithful servants" (Psalm 116:15). The very purpose of God's action in the world—the denouement of the Big Story—is to "swallow up death forever" and "wipe away the tears from all faces" (Isaiah 25:8). A too-narrow definition of pro-life abdicates our God-given calling to continually support and nurture life at all of its precious stages. For so many in the Christian community, including myself for much of my life, it obscures the moral weight of other life issues that also demand our attention. This kind of absolutist ethical schema may be psychologically comforting, but it does a disservice to Scripture's full-throated witness to the equal dignity and value of all image-bearers of God.

What, then, might a more holistic pro-life ethic look like in a warming world? It would continue to fight for a world where abortions are exceedingly rare[29] while also taking seriously the deadly consequences of climate change and pollution. The annual infernos engulfing the American West, the storms regularly battering the Gulf Coast, and the record-breaking floods bankrupting Midwest farmers year after year would be rightly seen as the threats to human life that they are. The standard industry practices that are releasing devastating neurotoxins and other pollutants into our air and water through the harvesting, processing, and burning of fossil fuels would be resisted. The death and disease spreading unequally across the globe would rouse to action all those concerned with protecting and defending life. A holistic pro-life posture in a warming world would recognize the threats that pollution and climate change pose to human life around the world and would fight for policies to slow its progress and to assist as many people as possible as they adapt to its impacts.

As our faith drives us to protect and defend life at all stages from the threats posed by climate change and pollution, the argument that there is only one faithful way to bring our defense of life into the public square will be exposed for the lie that it is. The arbitrarily partisan battle lines drawn by those in power will fade away as a radically integrated public discipleship takes shape.

We will defend human rights as necessary prerequisites to defending the inherent worth and dignity of all people. We will take seriously the scourge of racial injustice and White supremacy in America as violent assaults on the image of God. We will defend the religious freedom of all faiths and of those who choose to practice no particular faith. Even if some may maintain sincerely held disagreements about same-sex sexual activity, a holistic pro-life ethic would nevertheless drive us to defend the *lives* of LGBTQ+ people against violence and would fight for their equal protection against abuse under the law, as Pope Francis did in 2020.[30] We would resist the death penalty, especially its unequal and racist application.[31] We would support alternatives to euthanasia, like universal access to palliative and hospice care.

And as our concern for life in the public square expands beyond the myopic bounds of abortion alone, a wonderful feeling will steal over the many Christians who agonize every few years over whether or not to vote for candidates who are manifestly antithetical to much of what it means to follow Jesus, but who say the right things about abortion.

They will begin to feel free. They will be free from the cynicism of one party that takes their votes election after election yet does little to defend against the full spectrum of threats aligned against God's gift of life.

They will escape the disdain of another party that views their sincerely held belief in the value of unborn life as hopelessly provincial. They can release themselves from the obligations of unwavering partisan loyalties and integrate a truly holistic ethic of human life into an independent Christian citizenship that answers to Jesus alone.

It will no doubt feel scary for many to leave the solid ground of moral certitude and to venture out into the murky waters of moral discernment, but there is freedom in these waters. There is freedom to use our God-given intellect to weigh the fullness of all policies that impact the ability of our neighbors, and all of creation, to flourish and thrive. There is freedom to bring Jesus' promise of "abundant life" (John 10:10) into every aspect of public life for the sake of God's glory, our neighbor's good, and the flourishing of all creation.

6

A STORY CAN CHANGE THE WORLD

THE CAR RATTLED as it inched its way forward before grinding to a stop. The air conditioner blasted cool air and exhaust fumes in equal measure, and only partially succeeded in drowning out the cacophony of car horns all around them. Robert drummed his fingers on the steering wheel nervously as he glanced up at the darkening sky again. From the back seat, moans of pain reached him, each one a fresh stab of worry and guilt.

They had been on the interstate for hours and still weren't outside the city limits. By order of the governor, every lane was now northbound. That didn't prevent each one of them from being transformed into a slow-motion conveyor belt, and Robert was becoming increasingly convinced they led nowhere.

The predictions had been growing more dire by the day as the forecast models honed in on the city. As a New Orleans native, Robert had lived through plenty of hurricanes. He was used to the usually placid waves of the Gulf of Mexico making their occasional intrusion into his city, only to recede again to rest within their usual boundaries. This time felt different though. The warnings from local meteorologists and elected officials had an edge of fear to them. The predicted storm surge was higher than usual, and everyone knew how tenuous the ill-maintained levees surrounding much of the city were. What was more, Robert's mother was older and frailer than ever. Having ridden out her own share of hurricanes, she was no longer equipped to face the days of uncertainty, power outages, and shortened supplies that often came after a storm.

So, Robert decided to get his family out. He had loaded up the family car, strapped his granddaughter in her car seat, and made his elderly mother as comfortable as possible in the back seat before easing his way onto I-10. He met a wall of cars, each filled with families as desperate as Robert's to escape the impending storm.

The moans from the back seat grew louder as Robert's mother's pain spiked. Hours in the same cramped position had strained her frail body to its breaking point. With miles of cars still in front of him, Robert made his way to the shoulder and turned around, joining the trickle of other cars moving against the flood of vehicles still inching their way north. They made it back home in a matter of minutes, and Robert immediately set about preparing as best he could for what was coming. He put towels under doors, tied down whatever he could, and barricaded windows. He and his family settled in for a restless night, comforting each other with stories of hurricanes past and praying against disaster. The wind steadily gathered strength outside while rain began to patter against their windows, disrupting their uneasy sleep.

Then, in the early hours of the morning, disaster struck. Robert was shaken awake by his three-year-old granddaughter. "Papa, papa, wake up," she urged. "There's water in the house." Robert hurried out of his bed and ran to the back door. Sure enough, a few inches of water were lapping underneath the threshold. For a moment, Robert let himself hope that perhaps that would be the worst of it. But the levees protecting Robert's and his neighbors' homes were already straining against the historic storm surge being ushered ashore by Hurricane Katrina, and they wouldn't hold much longer.

As the water began to pour into the home in earnest, Robert's instincts kicked in. He shouted throughout the house, going from room to room to rouse his sleeping family. They climbed the stairs to the second floor as the waters lapped at their heels. Reaching the top floor landing with nowhere else to go, Robert led his family up into the attic to try to escape the inexorable march of Katrina's waves, at this point chasing them upward at a rate of more than one foot per minute. They huddled together in the stifling space, shock crashing over them at the speed of the rising waters. Minutes ago, they had been asleep in their beds. Now their beds were underwater.

They experienced a few brief moments of reprieve before water began bubbling up through the gaps in the attic's trap door and between the plywood subfloor and the eaves of the roof. With no higher ground to run to, Robert cast wildly about for anything useful that might be hidden away among the souvenirs of forgotten years—macaroni art and trophies from competitions long since over. Then his eyes fastened on it: the ax leaning lazily against the attic corner. For most of its life, it had merely chopped the occasional piece of firewood or bitten back the particularly intrepid invaders to the landscaping. Now it would be the tool of their salvation.

Robert gripped it tightly in his shaking hands and swung it as hard as he could. With a dull thud, it connected with the two-by-fours above his head. Methodically, Robert opened the hole in his roof as the waters rose around him. When water began falling onto his upturned face to mingle with that swirling around his knees, he knew he had done it. He boosted his family through the escape hatch, but his relief was short-lived. He had led his family from one danger straight into the mouth of another. They had escaped the rising water but now were huddled together on their roof against the full force and fury of Hurricane Katrina swirling around them. They watched as their neighbors' homes were lifted off their foundations and swept into the river that had replaced their street. They watched them drift away and then break apart against tree branches and telephone poles.

He didn't know how long they sat there, hugging each other and praying for the storm to stop. The next thing Robert remembers, their perch began to tilt and sway. Slowly, their home too slipped into the current that surrounded them on all sides. As the roof began to buck and tip, the family held onto each other for dear life. They gripped loose shingles and gutters—anything their hands could find to anchor them to their sloped life raft. Robert's mother, her grip slackened by age and her energy flagging from the family's hours-long effort to evacuate the day before, lost contact with the roof and was thrown into the frothing waves. Robert and his family scrambled after her, reaching desperately into the churning water for an arm or a leg. Miraculously, the family was able to pull her, shivering and terrified, back up onto the roof.

As the family regrouped and redoubled their efforts to stay firmly planted on the roof, an unbearable realization rippled through them. One of their number was missing. In the frantic effort to rescue the oldest member of the family, the youngest had fallen off the roof too— the granddaughter who had awoken Robert, warned the family of the impending danger, and given them a chance to escape it. Days before she may have been riding her tricycle down her sun-drenched street or reveling in the sticky joy of letting a popsicle slowly melt down her hand, but now the family's three-year-old savior had slipped off the roof and been carried away. They would never recover her body. She would join the more than twelve hundred estimated souls that were claimed by Katrina's diabolical power.

A few days later, having had no blanket or any other extra way to warm herself after her own fall into the churning water, Robert's mother would succumb to pneumonia and join her great-granddaughter once more.

Stories Can Change the World

Stories have power. As we explored in chapter three, words have the power to create worlds, to shape reality, to define the horizon of the possible. When Robert told me his story on a warm New Orleans summer evening in 2015, my world changed forever.

Each of us has a story to tell too. Perhaps it is a story about how God is working in our lives to bring about a deeper appreciation for his world. Or maybe it's a story about how the Holy Spirit just wouldn't leave us alone until we looked the injustice of climate change full in the face. We might have a story about a cousin with asthma, an uncle whose farm is going belly up, family in California who can never seem to outrun the flames, or friends from the Gulf Coast who hold their breath every year as they get battered by storms that just seem to get worse and worse.

We all have a story about why climate change matters to us because we're all human beings living in the world. Evangelical climate scientist Dr. Katharine Hayhoe often says that everyone has a reason to care about climate change—they just don't always know it yet. If you're breathing oxygen, drinking water, eating food, and love other humans that are doing the same, then you have a story about why climate change matters to you.

As Christians, our stories have a unique accent. We're not only concerned about climate change because of the ways it impacts us. We're concerned because the Big Story makes it perfectly plain that God loves his creation and expects us to love and care for it too. We're concerned because, like our God, we revere and honor the gift of life and want to restrain the death-dealing forces unleashed by climate change. We're concerned because we're passionate about enacting the good news of the kingdom of God right here, right now. We're concerned because we don't know how to obey Jesus' command to love God and to love our neighbor in a warming world without addressing the threats that climate change poses to our neighbors and God's creation alike.

We all have a story inside of us—a story with the power to subvert expectations, to shatter stereotypes, to move those in power to action, to convict a friend, to encourage other storytellers. Each of us has a story with the power to change the world.

Robert has made it his mission to tell his story to as many people as possible. He shares it at incredible personal cost, forcing himself to relive the worst experience of his life over and over again in the midst of total strangers. He shares it in the hopes that it just might make some difference for his community and for people he doesn't even know who might be spared the pain he and his family were forced to endure. I doubt any of us can claim this level of fortitude when we tell our own climate stories, but sharing our climate stories in the church does often require a certain kind of courage.

There is an extra layer of risk and grief for those of us called to address climate change in the church because sharing our stories carries the real possibility of ostracization and estrangement from friends, family, and church communities. I've heard more stories than I can count from young Christians who have been made to grieve not only for the heartbreaking consequences of unmitigated climate change but also for the strained relationships that their climate concern has wrought. When children raised in a church stand up to call that church to enact the values they have been taught within its walls by doing something about climate change and are met with gaslighting and indifference, spiritual and relational damage is done.

I am intimately familiar with how it feels to make the mental calculations in order to balance courage and integrity with the particular and painful risks of this unique kind of spiritual alienation. Let me be clear: these risks are real, and they are different for everyone. However, as real as these perils are and as necessary as it is for each of us to calculate our ability to withstand them, I believe that when we choose to keep our stories to ourselves, almost all of us are miscalculating.

A 2021 survey by the Yale Program on Climate Change Communication and George Mason University tells us that 64 percent of Americans are concerned about climate change and that 67 percent of Americans say that climate change is personally important to them. Yet, only 33 percent of Americans say they discuss climate change "occasionally" or "often" with friends and family.[1] And the worst part, according to a 2016 Yale study, is that among those who are either "interested" in global warming or think it is "important," more than half of them are neither hearing it discussed nor discussing it themselves with any regularity.[2] Most of us care about climate change, but we don't know that because most of us aren't talking about it. Researchers have coined this the "climate spiral of silence."[3]

At the same time, some of the same researchers have found that friends and family are some of the most trusted sources for climate change information[4] and that discussions of climate change within close social circles can lead to greater understanding and acceptance of key climate science facts.[5] One study even found that the conversations that yielded the most statistically significant increase of global-warming acceptance was when daughters spoke to their conservative fathers about climate change.[6] In short, we're shortchanging the power of our stories.

I think this is especially true in the church, where the stories we discussed in chapter two conspire to accelerate this spiral of silence. The explicit and implicit hostility so many of us feel in our churches toward conversations about climate change keep us quiet lest we suffer the spiritual trauma of isolation from our spiritual community. If the Yale figures cited above can be extended to our worshiping communities, though, then the chances are good that there are at least some members of a given church who are personally concerned about climate change and that those who aren't yet concerned could take an interest if a member or members of their community found the courage to speak up. Ironically, in our attempts to avoid isolation, many of us stay silent and end up isolated from other members of our church who share our climate concern.

Whenever I give a talk or presentation and I get to the "what you can do" portion, I always start with "talk about it." I know full well that it's not the most exciting way to respond to the climate crisis. There are many other ways, and we'll get to those in later chapters. I want to talk about stories first, though, because I'm convinced that our own stories are the most powerful tool that we have in our toolbox to respond to the threat of a changing climate.

I believe this not only because the research bears it out but because I have experienced it to be true in my own life. I have been transformed by climate stories: my brother's story after his return from New Zealand, Larry's story of his love for Kayford and of the intimidation he faced because of it, Margaret's story of capricious rains and unfamiliar dance steps, and Robert's story of loss and resilience. I carry each of these stories like a precious gift, and each of them has shaped my own climate story in profound and lasting ways.

So how do we talk about it? How do we share our climate stories in ways that convict, invite, and break down walls?

The Messenger Is More Important Than the Message

One of the most important truths about climate-change communication is that the messenger is almost always more important than the message. That is to say, it matters less *what* someone is saying, and it matters a great deal more *who* is saying it. This becomes eminently obvious to any of us who have ever heard an extended family member say something like, "Global warming? I don't know much about it, but if Al Gore says it's true, then I'm not going to believe it!"

These sorts of comments can feel discouraging, but they are actually more encouraging than they first appear. The person who makes this kind of comment is essentially saying that the only good reason they have for disbelieving what the science is telling us about climate change is because Al Gore is the only one telling them about it. This means that they might believe it if someone else tells them about it—especially someone they know, love, and trust. You may be thinking, *But I don't know as much about climate change as Al Gore. I'll never communicate like he does.* All I have to say is, "Promise?"

With all due respect to Al Gore (and he does deserve it), he will not move many American Christians who are currently on the fence about climate change. Is it because he doesn't know enough of the science? Is it because his slideshows aren't state of the art? Of course not! He knows the science better than almost anyone on the face of the earth, and his slideshows are impeccable. It's because polarization and hyperpartisanship mean he'll never be the right messenger for them. He's not going to be able to reach your Uncle Joe or your grandma. But do you know who just might? You.

Be Curious

I had a professor in seminary who used to tell us that the task of pastoral care is to remain curious. I think it just might be the task for all Christians, full stop. In today's hyperpolarized world, it feels like we're always just sorting each other into binary categories—D or R, pro or anti, us or them. My friend and teacher Randy Woodley would say that this is a symptom of the Platonic dualism we discussed back in chapter two. For many of us in the West, binary thinking is our epistemological inheritance thanks to movements like the Enlightenment, when empiricism sought to categorize and taxonomize everything it could lay its hands on, including human beings.

This means that if we are to be people who resist Platonic dualism's reflex to flatten out the complexity of the world around us—including our fellow humans in all their rich and textured humanness—then we need to practice the art of curiosity. Curiosity is the antidote to dualism's legacy of polarization and dehumanization.

As we tell our stories to those closest to us, we must be just as ready to receive their stories. We must listen—truly listen—as much as or more than we speak. We must practice active listening, which means listening to truly hear and not simply pretending to listen as you formulate your next talking point. We do this not merely as a tactic to win their trust but out of a genuine desire to experience them in their full humanity. We listen to complicate our understanding of who they are, and to resist the trap of dualistic categorization. In so doing, we can humanize conversations about climate change, which have trafficked in dehumanization for far too long.

Bond, Connect, Inspire, Invite

Once we've recognized our power as effective messengers for our friends, family, and church community and have committed ourselves to the art of curiosity, how do we actually connect with people about climate change in a way that might make a difference? I get this question all the time. So does Dr. Hayhoe. She offers a formula for constructive conversations that I think is helpful: find something you both care about, connect the dots to how climate change is already affecting what you both care about, and share a constructive solution that you can both agree on. She simplifies this into a maxim: bond, connect, inspire.[7]

I love this framework because it recognizes the importance of the messenger and it requires curiosity. However, as important and useful as Dr. Hayhoe's framework is, I have to admit that I'm feeling a little impatient with merely hoping for constructive conversations. Constructive conversations are important but can never be the end goal of sharing our stories. The ways in which we share our stories should always land at an invitation to action. This invitation can be small, like inviting them to consider what their own climate story might be. It can be bigger, like asking them to join you at a march or a sit-in. Whatever the ask, the point is to offer an invitation for others to imagine themselves in the story of action as a positive part of the solution. In order for our stories to be invitational, we must do not three but four things: bond, connect, inspire, *and invite.*

The good news about finding common ground and connecting climate change to something you and your conversation partner already care about is that it is easier than you think. You don't need graphs and data points. You don't need the latest articles from *Nature* and *Science* in your back pocket. You don't need to be an expert on climate science at all. The only thing you need to be an expert in is your own story— your own experience of how God has led you to understand that doing something about climate change is part of what it means to follow Jesus in the twenty-first century. What you know well are the people who have moved and influenced you, the books that have caused you to think differently, the films that have enlightened you, the sermons that have convicted you, the experiences that have shaped you, the worship songs that have moved you. These are the stories that those closest to you need to hear, not more science and statistics.

The truth is that the science is so settled and so readily accessible that most of the people who are going to be moved by the science have already been moved. The rest need to hear stories that connect with their sense of self and the values they hold dear. And they need to hear this from the people who are closest to them. It is much easier to discount an esoteric scientific study written by a nameless, faceless scientist than it is to discredit the personal story of transformation from someone standing right in front of them who is known and loved and trusted— someone like you.

Instead of hard data, we do best when we focus on the emotions of climate change—sharing how climate change makes us feel, and how it makes us feel to be doing something about it. It can be comforting to hear someone be vulnerable about how scared climate change may make them feel. Most of us are scared; we're just not telling each other. And then it can be transformative to hear how someone is converting that fear into hope by taking positive action. And when we share these emotions, we are most effective when we connect them to the things that matter to the person with whom we're sharing our story.

How do we know what's important to them? We learn that when we practice curious, active listening. Listen long enough, and someone will always tell you what's important to them.

Perhaps your sister is a real mama bear who gets fired up at anything that might pose a threat to her kids. You can bond over your shared love of her children—your nieces and nephews—and why you are scared for their future. Maybe you share a love for hunting and fishing with your skeptical uncle. You can share how surprised you were to learn that habitat loss and land-use change threaten both our climate and the populations of the animals you both love to hunt. If you're like me, you have extended family who are active in the Right to Life movement. You can share how concerned you are for all the threats posed to unborn lives by fossil fuel pollution and climate change and how you understand climate action as an expression of the pro-life convictions they hold so dear.

Some may think what I'm proposing is cynical manipulation, but I think it's closer to hospitality. It's sharing your story in such a way as to invite those receiving it to imagine themselves in the story of action along with you.

Once you have connected climate change to shared values, move quickly to solutions. People don't want to be dragged down into a pit of existential despair and left there. They need to know there's a way to climb out. Sharing solutions assures them that there is a way out. Dr. Hayhoe likes to say that the fact that climate change is caused by humans is actually good news. If it were the result of some force outside of our control, then there would be nothing we could do. But it's not. Because we are causing the problem, we are also able to find the solutions we need to address it.

And, thankfully, there are a lot of solutions! Electric vehicles are significantly less carbon-intensive to manufacture and power—even when the electricity used to run them is generated with fossil fuels—and costs are already on par with internal-combustion engines. When all subsidies are taken away, large-scale renewable energy like wind and solar (think big solar and wind farms rather than solar panels on roofs) are already cheaper than coal in almost every US market and have reached cost parity with natural gas in most of the country too.[8]

Lithium-ion batteries always made good sense for electric vehicles but aren't up to the task of storing power generated at utility (i.e., power plant) scale. Cutting-edge battery storage technology is now flipping the script that once said solar and wind are only viable when the sun shines and the wind blows. Emerging iron-air batteries use rust (you read that right—rust) to generate more than a hundred hours of continuous power. What's more, they last for up to thirty years and use abundantly available and nontoxic materials (iron ore), all at one-tenth the cost of lithium-ion batteries. Storing the energy from sunny and windy days to be used on cloudy and calm days was always a major missing link in our technological capacity to jump from fossil fuels to renewable energy. With these kinds of breakthroughs in storage technology, a 100 percent clean, renewable energy grid is not only possible but only a matter of time.

Investing in public transit can clean up toxic air by taking cars off the road, as well as make transportation more accessible and equitable for all people. Regenerative agriculture can harness the ancient power of soil to lock away massive amounts of carbon and to fix it in the ground where it can be put to use creating rich, nutritious food. Market-based approaches like a tax on carbon or a carbon border adjustment can make sure the true costs of fossil fuels are paid by those who produce it rather than being passed on in higher health care costs, higher insurance premiums, and increased mortality rates for millions around the world.

Next generation small modular nuclear reactors (SMRs) have the potential to replace the hulking, centralized behemoth nuclear plants of the past with streamlined technology at a fraction of the size and price. Silviculture and new methods of forest management can take into account both ancient human wisdom and advances in modern science to keep forests healthy so they can suck as much carbon dioxide out of the atmosphere as possible. Food co-ops and mutual aid can ensure that nobody needs to go hungry even as the effects of climate change continue to be felt around the world.

I could go on. The list is almost endless. Most people are surprised to learn that we actually know how to solve climate change. We have the knowledge, and we have the technology. All that's missing is the will. Our stories can help create that collective will, one person at a time, to finally do what needs to be done.

After you've shown them the way out of the pit of the problem by offering solutions that connect to your shared values, they need a rope to get from the bottom to the top. They need toeholds to get them started on their ascent, and they need lots of encouragement and support to get them all the way out. That's why we should always be inviting people to do something tangible to respond to our stories. People need to be invited to act in ways that feel commensurate with the scale of the problem we face. And they need to feel like they are doing it with other people who are like them. This is what social scientists call a "positive social norm." In the US church, this is especially crucial because, as we've seen, many Christians see relatively little evidence that other Christians are concerned about climate change and are doing something about it. But that is beginning to change, and the change accelerates every time we tell our stories, take action, and invite others to take action with us.

Importantly, this formula of connecting shared values to climate change, offering solutions, and inviting to action works for people at all stages of climate-change belief. A new kind of climate-change denialism is being recognized by researchers, particularly among younger generations. These young people do not doubt that climate change is happening. They feel acutely the severity of the climate crisis—and that is precisely the problem. A creeping dread and sense of doom is overtaking many, convincing them that the apocalypse is inevitable and that nothing can be done. Many are even forswearing having children, convinced it is unethical to bring an innocent life into such a mess. These people need connection, solutions, and invitations too! After all, the antidote to despair is not hope but action. We can't will ourselves to have enough hope before we can start taking action. The formula works in the opposite direction. We find hope when we take action. Especially when we act together.

How to Tell a Good Story

On a rainy Parisian day, I found myself sitting inside an empty church with a handful of other Christian climate organizers and activists who were also in Paris for the COP21 United Nations Climate Change Conference. They were heady days filled with high-level meetings, celebrity sightings (if John Kerry and Ban Ki-Moon can be considered celebrities), and overwhelming optimism. There was an irrepressible sense that we were finally going to do it. We were finally going to achieve global consensus, break the gridlock, and do what needed to be done to avert climate catastrophe.

Yet, there was a nagging doubt—a doubt that I and the dozen or so others gathered inside the church knew intimately. While the world's leaders may have gathered together and appeared to be mostly on the same page, the world's population was not. In particular, the populations of a few of the world's wealthiest developed countries—like Canada, Australia, the United Kingdom, and the United States, where climate change remained one of the most polarizing issues in society—remained badly divided on how, or even if, to respond to climate change.

And so it was that my friends and I, all working in some way or another to reach the Western church with the message of gospel-rooted climate action, gathered to learn how we could respond to the disconnect between the negotiations happening in Paris and the doubts we would meet once we returned home. There to teach us that day was George Marshall, a social science researcher based in the United Kingdom. Marshall had just completed a research project exploring what message frames and tactics work best to move conservative faith communities. I wish I could re-create everything that Marshall shared that day (although his book, *Don't Even Think About It: Why Our Brains Are Wired to Ignore Climate Change*, is a good substitute). The data I found most compelling had to do with his research into the marks of what made for a good climate storyteller with this particular audience.

Bad storytellers, said Marshall, ignore the unique identities and values of the people to whom they are telling their stories. Our identities are what anchor us within our communities. They give us a sense of belonging and provide our lives with meaning. Our values are what drive our actions within our communities of belonging. Taken together, connecting with someone's identity and values is one of the most powerful ways we can motivate people toward sustainable, positive action.

Bad storytellers don't care about the identity or values of the person they're talking to. They simply focus on their own. They talk about how climate action affirms *their* sense of self and how their commitment to *their own* values are deepened by taking action, without ever acknowledging the values of their interlocutor.

The result, says Marshall, is that someone listening to a bad storyteller will hear something like this:

Here are all the reasons that you and your loved ones are wrong about climate change. Here are all the reasons *I* and others who are nothing like you are right. Here are all the ways that you need to radically change the life that you love if you want to be right like *me*. Doing so will alienate you from every person you know and love, but don't worry, because it'll make you more like *me*, and the world will be more like *I* want it to be.

Compelling stuff, right? I mean, sign me up.

In order to avoid being a bad storyteller, says Marshall, we need to spend our time connecting climate action not to our own identities and values but to the identities and values of the people we're trying to reach. For many of us, this will mean only slight adjustments. If we are trying to reach our friends and family, chances are that we already share a common identity and hold many of the same values. It's simply a matter of helping those family and friends understand that our climate concern stems from these shared identities and values.

When we are able to tell our stories in ways that connect to the identity and the values of those we're trying to reach, says Marshall, the listener hears something like this:

Here are all the things that *you* care about and that make your community great. Here are all the people just like you, who care about the things that *you* care about, who are taking action to address climate change. When you join them, you deepen your connections to these people, and the world becomes more like *you* want it to be.[9]

Putting It All Together

Thanksgiving is approaching. Like every year, Kevin's extended family will gather for Thanksgiving dinner. And like every year, Kevin's Uncle Bob will regale the table with his—shall we say "charming"—perspectives. Kevin knows from past Thanksgiving addresses that his Uncle Bob doesn't think much of climate change. It's also quite clear, though, that he doesn't have much depth to his critique and is merely parroting talking points he's heard from cable news and from others around him.

Kevin has a good relationship with his uncle. He used to take Kevin hiking when he was younger, and he introduced Kevin to fly fishing for the first time—a sport he and his uncle share a love for and that they still do together to this day.

Kevin doesn't share his own views about climate change with his Uncle Bob. It just hasn't seemed worth it. But recently Kevin took a course on watershed management and learned just how vulnerable to climate change are the waterways where he and his uncle love to fish. Plus, the other day he ran across an article connecting the declining salmon populations in their area to warmer ocean and river temperatures—warmer temperatures driven by climate change.

So, after dinner Kevin decides to pull his uncle aside for a conversation.

"Uncle Bob, we going fly fishing this spring?"

"You know it, kid! That is, if there are any salmon running this year. Seems like there are fewer and fewer every time I go out there."

"Yeah, I've noticed that too. It seems like there are fewer even since I was a little kid. You know, I read an article recently about how climate change can threaten salmon populations."

"That's a bunch a baloney, kid. I don't buy any of that climate change nonsense."

"I know, Uncle Bob. I didn't used to either. But then I learned that the science is pretty solid, and that 97 percent of climate scientists agree that climate change is happening and that we're the main reason why."

"97 percent, huh?"

"I used to want to find another explanation for all the weird weather we're seeing and for why the salmon stock is dropping. I didn't want to be responsible, you know? But I've come to see how the fact that all these things are caused by us is really good news."

"Good news? How do you work that out?"

"Because it means we can do something about it! If the salmon were dying off for no reason, it would be terrifying. But because we know why these things are happening, we can find solutions. And there actually are a lot of solutions!"

"Yeah, like raising taxes, killing the economy, and the government taking away my hamburgers . . ."

"Not exactly, Uncle Bob. Something I learned in my class this semester is that salmon are naturally really adaptable. Otherwise, they wouldn't still be around! But they're having a hard time adapting to climate change because we've already stressed them out with all kinds of other things, like dams making it harder for them to migrate upstream to spawn and pollution runoff that's making streams and rivers harder for them to live in. While we absolutely need to address the root causes of climate change, we can also address some of the contributing factors that are stressing the salmon out so much."

"Like what?"

"Well, I just learned about this group in town called Save Our Salmon, and they work to keep pollution out of the river by planting native species all over the watershed. These native plants are good at capturing runoff from farms and roads around the river and filtering out the toxins. When there are lots of native plants in a watershed, the rainwater that eventually reaches rivers and streams is much cleaner and safer for everything living in it—including salmon! Save Our Salmon is hosting a river clean-up and native-planting event this spring before the big summer salmon runs. A bunch of people have already signed up on the Facebook event. I saw Pastor Bill is going to be there. Do you want to do it with me? We can bring our poles along and see what we can catch afterward!"

"Pastor Bill, huh? He's a good man. You know that's not a bad idea, kid. I'm in."

"Great! You know . . . there are also some protests being planned against a proposed oil and gas pipeline whose route runs right over the river too. If it spills, it could spell disaster for the salmon—not to mention all of us who get our drinking water from the river too. I was thinking about going to that too. Would you be interested?"

"Protest! I ain't no antifa Commie, kid."

"Alright, Uncle Bob. We'll work up to it then."

~

Okay, so that conversation was a little corny, but it does help us get a picture of what the principles I've shared above can look like in the real world. Kevin found something that he and his uncle both cared about—fly fishing—and he connected it to climate change. He shared how the declining salmon population made him feel and how it made him feel to learn that there were solutions. Then, he not only offered a solution, but he invited his uncle to be a part of the solution with him. He contributed to a positive social norm by letting his Uncle Bob know that people like him—his nephew, Pastor Bill—were taking action too. Using George Marshall's framework, Kevin affirmed Uncle Bob's values and identity, he showed him all the people like him who were already acting, and he helped his Uncle Bob imagine how taking action will contribute to a world that Uncle Bob wants to live in—a world filled with salmon!

Now, you may still be thinking, *That's not how my uncle would respond!* And I get it. This interaction is, by definition, contrived. Many of us have had conversations that have looked quite different from this one—conversations that were doomed from the start, where we try our best to share our knowledge and our heart, and we just don't get anywhere.

There is, of course, every possibility that Uncle Bob will not listen to Kevin, that he will not be open to attending the river clean-up project. There is every possibility that he will throw names around, cast aspersions on Kevin's character, and even question his faith. He may even forget to invite him to go fly fishing next spring. If you share your climate story enough times, it's bound to happen. It has to me plenty of times, and to many others.

This is part of the unique pain we experience as Christians compelled to call the church to deeper faithfulness on climate change. If we are to be faithful in responding to Christ's call to proclaim the good news of the coming kingdom of God to all creation, we cannot avoid calling our fellow Christians to take climate change seriously. We may lose relationships. We may lose credibility with friends and family. We may be on the receiving end of some nasty social media comments.

I can't promise that the principles outlined above will protect you from difficult, even painful, interactions. But I believe with all my heart that if we do our best to keep trying to be the best communicators and storytellers we can be; if we keep inviting people into action with us; if we keep connecting climate change to people's hearts, identities, and values; if we refuse to acquiesce to the false perception that Christians don't care about climate change and instead create a new narrative that bold climate action is simply part of what it means to follow after Jesus—then we might just find that we have more power to effect change than we ever dreamed possible.

The Story That Changed the World

Though the stitches in their sides seared, the women ran on. Fear and exhilaration filled them in equal measure. Could it be true?

The predawn light filtered through the trees as the sun began to rise in earnest. Their feet caught on rocks and the hems of their robes, causing them to stumble, but still they ran faster. They must reach them—his friends. They must tell them.

Even as they ran, doubts chased them like bloodhounds on their scent. Their common sense was returning as the initial rush of adrenaline began to wane. Who would believe such a fantastic story? Surely, they would face questions: How could they be sure of what they saw? Could they prove it? Were there any men there who could corroborate it? They were used to hearing these kinds of interrogations, used to having holes poked in their perspectives, their opinions going unnoticed. They were women, after all.

And still, in spite of their doubts, they ran. The quiet of the garden began giving way to the bustle of city streets. Vendors were setting up, preparing for a busy day ahead. The first shoppers of the day were venturing out of their homes to make their way to the market, eager to replenish household stores depleted by the events of recent days. Heads turned as the women streaked past. They shouted hurried apologies over their shoulders to the scandalized bystanders in their wake.

Finally, they stopped in front of the house they had been so desperate to reach. Gasping for breath, they looked one to another, finding in the faces looking back at them the same mixture of eagerness and apprehension they each felt. After another moment's pause, the woman nearest the door raised her hand. Her fist pounded in the same hard staccato that her heart was hammering against her chest. *No going back now.*

After a few moments—or maybe it was an eternity—the door opened a crack. A man's face, barely visible through the slit between the door and the frame, peered out cautiously. The woman nearest the door found her voice first, "Peter. Let us in. You're not going to believe what's happened."

7

GOD'S PLEASURE, OUR JOY

BY MY SOPHOMORE YEAR OF COLLEGE, I had a decision to make. I was now three years out from my brother's announcement to the family about his own epiphany and subsequent transformation. In the intervening time, we had had countless conversations about *his* journey, the arguments that had moved *him* to action, and the resources that had deepened *his* knowledge and understanding. And I had begun to have my own experiences that were sharpening my sense of calling to address the brokenness of climate change with my own life.

It was time for me to begin writing my own story of action.

Expect Joy

And so, I got started. I began to look for opportunities to align my values with my deepening knowledge. Shame at my staggering ignorance and profound guilt for my complicity in the death-dealing status quo precipitated sweeping changes. I joined the student organization on my campus dedicated to environmental stewardship. I started cutting out meat from my diet, one meal at a time. When I moved off campus the next year, food scraps were composted, vegetable gardens were sown, and buses and bikes were ridden while my car sat in the driveway.[1]

For a time, I was able to sustain these changes with sheer force of will, but my efforts were untethered to anything deep or abiding. They were acts of self-flagellation enacted through gritted teeth and out of overwhelming obligation. My resolve soon began to slacken as burnout licked at the edges of my will.

Maybe you can relate. Climate change is such an overwhelming challenge, with more than enough guilt, shame, and blame to go around. Even a basic understanding of the physics of our changing climate and how that cold calculus expresses itself in deeply unjust ways around the world is enough to plunge an averagely empathetic person into despair, anger, and fear.

And why not? When coming face to face with the gravity of climate change, these darker emotions live right at the surface. They rub us raw with their grating intensity. For some of us, they are our daily companions.

What else is there to feel but anxious about living through the planet's sixth mass extinction?[2] What else can we feel but dread as homes are lapped by rising tides, livelihoods are crumbled to ash and carried away on Santa Ana winds, and children are poisoned by noxious air and toxic water? What else can we feel but shame as the land God has commanded us to hold in trust for our children and grandchildren (Leviticus 25:23-24) is desecrated? What else is there to feel but guilt at our own ignorance and complicity, and rage at the behemoth vested interests who have worked so hard for so long to keep us in the dark about all of it?

It is important to name and honor these emotions. Yet, as valid and human as these emotions are, they will do little to win a safer and more habitable future for ourselves and future generations. Social scientists have reported for years that while a bit of justified anger can spark someone to action, it quickly consumes itself.[3] For sustained, long-term action—the kind needed to face down the threat of climate change—we need more than anger. Every one of us needs a sense of agency. We need a community of belonging. We need hope that our actions can make a difference. We need joy.

Joy is a concrete rebuke to the death-dealing powers and principalities that traffic in fear and despair. It opens the aperture of our mind's eye to see beyond the headlines in front of us to the in-breaking of the new creation all around us. It is a winsome witness to the inevitable victory of justice, peace, and delight over sin, despair, and death. It grounds our action in a bedrock that runs deeper than the inevitable ebb and flow of long-term advocacy for climate justice. It binds us together in our common efforts. And, perhaps most importantly, it is just plain good for our souls.

But joy can be a hard sell. Among many in the climate movement, cynicism and a simmering rage are often worn as a badge of honor. They are coping mechanisms against the daily trauma of watching the beautiful, wondrous world die a little more each day—self-medication from, as author and naturalist Aldo Leopold put it, "liv[ing] alone in a world of wounds."[4] I'm certainly guilty as charged. Joy can feel extravagant, indecent. What right do I have to feel joy while entire communities lose their homes and people lose their lives? Joy feels like a betrayal—like the purview of the privileged and the ignorant alone.

But as I kept slogging my way through my sober rituals of obligation, I had to admit something to myself: I needed more than the guilt and the shame and the blame. I needed a vision for personal action that moved beyond dour duty and was rooted in something deeper. I needed joy.

The truth is, as all of us go about the work of aligning our lives with our Christian values in response to the realities of climate change, we all need joy. Otherwise, we simply won't be sustained for the journey. And once we give ourselves permission to look for joy in the midst of our climate efforts, it is everywhere around us, ripe for the picking.

My first taste of it came on a road trip to Washington, DC, that same sophomore year. I had joined the Environmental Stewardship Coalition on campus and had thrown myself into their efforts to raise awareness about the impacts of mountaintop-removal coal mining. I still wouldn't meet Larry Gibson for two more years, and it was largely the work I did with this student group that inspired me to lead the trip to meet him and others in West Virginia.

Part of the campaign in those early days included advocating with Congress for a bill that would prevent operators from poisoning local water supplies with their waste and would instead force them to remove it from the site completely and dispose of it safely. If the rule change were to be adopted, it would effectively make mountaintop-removal mining uneconomical overnight.

So it was that I found myself crammed into a twelve-passenger van, skipping class and hurtling down a predawn interstate, a watery early March sunrise just beginning to break over the horizon. I knew a few of my fellow truant sojourners, but most were new or recent acquaintances. Yet, we were united in our effort to lift up to our elected officials the voices of those being harmed by mountaintop removal.

It was my first real experience with direct congressional advocacy, and I was nervous. Who was I to speak to an elected member of Congress? To educate them? To chastise them, even? To request that they use their power to do something? I was a college student—twenty years old and almost entirely ignorant of the legislative process, but I took courage from my co-conspirators. I was bolstered by their willingness to also hop in a van with near strangers and drive halfway across the country to stand in a congressional office and prophesy against a moral injustice—to skip class and get behind on their homework for something bigger and more important. I was empowered by the leaders who hosted us in Washington, DC, that week as they reminded each of us that we had every right in the world to speak our minds with our elected officials for the simple reason that we were their constituents and that those powerful, intimidating men and women worked for us.

That week of learning shifted something in me. Together, we delighted in singing along to the radio for twelve hours straight, in giggle fits that just wouldn't quit in our sleeping bags on the hard church floor every night, and in finding the best hole-in-the-wall Chinese restaurant in the city. It exposed me to the possibility that maybe my atomized acts of penance could be something more. That week helped me see that when joined to the actions of others, in a community of belonging, they could be transformed into offerings of joy.

After getting a taste for the joy and hope that are possible in joining my actions to a community, I was hooked. It was like taking a long, cool drink of water after ages in the desert. I didn't even know how desperately thirsty I was.

I went looking for more of it—and I found it. I found it in plant-based kitchen experiments with housemates and with my future spouse, discovering new skills and appreciation for the gift of food at tables laden with flavor and love in equal measure. I found it in shared celebrations of those rare and exhilarating policy successes that our campus environmental group experienced together (like getting our Republican congressman to co-sponsor that mountaintop-removal bill), the precious products of our communal courage and conviction. After graduating, I would find it in the honor of getting to shape emerging leaders myself, shepherding new activists into the movement as I had been shepherded.

As I searched out more opportunities for joy in my climate action, I began to notice a curious alchemy at work. It occurred when I would plunge my hands into the earth or look my neighbors in the eye on the bus or cook new plant-based meals with friends. Guilt was slowly transforming into gratitude, despair into joy, isolation into community.

And why not? After all, when we take steps to align our actions with our values in community, we are living into our deepest identity as image bearers called to mirror the loving care of the Creator. As those creatures unique among created things, we were made in community ("male and female he created them," Genesis 1:27) and invited to speak God's loving words over creation together. We were blessed by God that we might then turn around and, together, bless all of creation as we point it toward its one, true king.

I was coming to understand my efforts as more than penance or obligation. They were beginning to look a whole lot like spiritual disciplines.

The Spiritual Disciplines of Climate Action

It may sound strange to speak of taking shorter showers or reducing our meat intake as spiritual disciplines. Isn't it a little over the top to put these seemingly quaint lifestyle practices into the same vaunted category as prayer, worship, and meditation on Scripture? Maybe, but maybe not.

Richard Foster describes spiritual disciplines variously in his classic book, *Celebration of Discipline*. At their core though, says Foster, the spiritual disciplines are about creating the conditions in which God "can work within us to transform us. By themselves," says Foster, "the spiritual disciplines can do nothing; they can only get us to a place where something can be done. They are God's means of grace."[5]

In other words, spiritual disciples are the means by which God forms us more fully into his people. They are the crucible in which our hearts are forged—ever so slowly—into hearts that more closely resemble the heart of God. If, as we've seen, God's heart is one that loves his creation and yearns for its liberation, then why wouldn't actions that cultivate this same love and yearning in us be considered spiritual disciplines?

Simplicity is a helpful case study. On its face, the eschewing of excess material goods may seem only vaguely related to following Jesus—especially to our modern capitalistic minds. Yet it is instructive that simplicity is so often included on the list of classical spiritual disciplines. Practitioners and teachers of the spiritual disciplines across time and space have recognized that the human heart, distorted by sin yet created for union with God, will grasp after anything that can give it a sense of center, rest, and wholeness. In societies and economies organized around ever-increasing consumption of goods and services, our feeble hearts seek to slake their thirst for communion by forming obsessive attachments with things rather than with the One who created all things.

This was true in first-century Palestine. That's why Jesus challenged his listeners on the Mount of Olives to choose which master they would serve—God or money (Matthew 6:24). For the same reason, he told the rich young man to break his attachment to his wealth by selling all he owned. He then warned all who were listening that it was easier for a camel to enter the eye of a needle than for a rich person to fully embrace the reality of the kingdom of God—not because those who are wealthy are inherently more sinful but because their sin-distorted hearts have so many more opportunities to form eternal attachments to that which isn't God (Mark 10:17-25).

This proclivity to form misguided attachments to our stuff was true in medieval Europe, too, when Francis of Assisi called his followers to make vows of poverty and to give away all their possessions as an act of liberation from their slavery to things. And it is most certainly true today, perhaps no more so than in modern industrialized countries like the United States, whose economies are predicated upon the proposition of infinite growth and unfettered consumption, whose powerful corporations have perfected the art of manipulating our feeble hearts through the manufacturing of desire, and whose people are awash in a sea of hypercommercialism, where the lines between advertisement and reality are almost too blurred to make out. Overflowing waste, soaring income inequality, and runaway climate change are enough to indicate where our hearts truly lie.

In other words, the spiritual discipline of simplicity does what spiritual disciplines do. It turns our hearts toward God and away from that which pretends to be, but is not, God. This is also how the field of virtue ethics speaks of habits and practices. Unlike deontological ethicists, who believe there is a universal set of moral laws to be followed by all people, or utilitarian ethicists, who believe something is ethical if its end result is to maximize the good for the most people, virtue ethicists believe that right living flows from one's affections. One's affections, in turn, are not hardwired before birth but are cultivated over time by one's practices. We've all heard the phrase "you are what you eat." Virtue ethicists believe "you are what you do." Or, in the words of one of the world's preeminent virtue ethicists, James K. A. Smith, "you are what you love."[6]

In one of his most celebrated books, *Desiring the Kingdom: Worship, Worldview, and Cultural Formation*, Smith tells a parable to help us grasp this concept. Martian anthropologists have journeyed to earth to study the religious practices of its inhabitants. They touch down in a metropolitan area of the United States, and they make their way to what they take to be a large temple. It has a massive parking lot to accommodate the thousands of pilgrims that throng to it every day. Inside, they notice banners with text and iconography to orient pilgrims to the space. Upon entering, they find a large map to aid the worshipers even further in their engagement with the various worship offerings of the temple. Large windows in the ceiling draw the eyes of the faithful heavenward while blank stretches of wall shut out distractions from the mundane world they have just left behind. Worshipers engage in a kind of ritual contemplation as they shuffle through the labyrinthine space, engaging spectacularly vivid icons meant to inspire worshipers to emulate the example of each saint that is depicted. This is a faith marked by beauty and a vision of the good life, not dour pietism or strict doctrine.

After observing these icons and other liturgical signposts, the alien anthropologists are drawn to observe what is actually occurring in one of the many smaller chapels. They are greeted by an attentive guide who orients them to the worshiping act and then allows them space to engage on their own terms. All around them are acolytes engaged in the liturgical practice colloquially termed "searching through the racks." Some are tentative and meandering, searching for something and unsure exactly of what but positive that they will find it here. Others are resolute, sure of that for which they search.

When they finally hit on a holy object, the worshipers proceed to the altar for the climax of the worshiping act. Behind the altar is the priest, who presides over the consummating transaction, the holy exchange of give and take. Blessed by the priest with a benediction, the worshipers leave the chapel with a holy relic, a means by which they might achieve the good life embodied in the icons that drew them into worship in the first place. Some leave the temple. Others continue to meander, to enter into another of the myriad worshiping spaces, and to enact the liturgy again.[7]

Perhaps you've figured out by now the true nature of the holy temple studied by Smith's Martian scholars: your friendly, neighborhood mall. And lest you assume Smith is merely being playful or exaggerating to make his point, he makes his intentions perfectly plain: "My goal is to try to make strange what is so familiar to us precisely in order to help us see what is at stake in formative practices that are part of the mall experience."[8] The Western, post-Enlightenment church, argues Smith, tends to privilege the head (i.e., doctrine, orthodoxy, right thinking) above all else. Human beings are understood to be, at root, believers who are governed by a set of conscious and unconscious beliefs. Battle-lines are drawn around theory, theology, and dogma.

As such, the power of practice (orthopraxy) to form our hearts and order our affections is often neglected. So, when Christians enter a mall (or our Amazon carts), we are often blind to the formative nature of the practices in which we engage. A religious approach focused on the head, says Smith, "is not really calibrated to see the quasi-liturgical practices at work in a site like the mall."[9]

I think something similar is at work in the church when it comes to climate change. It's so easy to get caught up in our heads—debating facts and statistics and studies—that we forget about the primitive power of practice. Battlelines get drawn around what you believe about climate change (never mind that a thermometer reads the same whether you "believe" in it or not). We are not merely brains on a stick, as Smith likes to remind us. We are embodied creatures, and the things we do with our bodies matter immensely. When it comes to engaging climate change in the church, we ignore the formative power of actions at our own risk.

Why else would Christians throughout history—in every time, place, and culture—worship regularly together? The very concept of liturgy is based on the assumption that specific things we do with our bodies in specific places at specific times with specific people form us in a specific way. Over the course of church history, it has been observed across time and space—with very few exceptions—that regularly worshiping God with a community of other believers is a necessary component of our sanctification—the Spirit's work of slowly, deliberately turning our hearts back toward God.

When we remember the power of practice to form our hearts toward particular loves, it suddenly doesn't seem so strange that intentional, regular practices to live in a more reciprocal relationship with creation might have the power to form our hearts in particular ways. Weeding in your garden every day from May to September might tune your attention to the complex world-within-a-world of microbe, earthworm, and dirt that we call soil. Turning the washer knob to cold every time you do a load of laundry might focus your buzzing mind, even for a split second, on our holy duty as image bearers of the Creator to do our best to mirror God's love and compassion to all of creation.

In chapter four, we observed that—according to the Gospels and Acts—the good news we are called to share is meant for the entire creation, is particularly good news for those on the margins of society, and is spread by the Holy Spirit working through us. If this is the case, then we should be pursuing those practices that form us in such a way as to expand our moral imaginations beyond human hearts alone, that move us closer to those who suffer, and that center the Holy Spirit and de-center ourselves. The cultivation of ecological virtue through simple, quotidian acts practiced over and over has the power not only to bring down the global temperature but also to form us to be more faithful heralds of the good news of Jesus Christ.

I'll never forget my own moment of epiphany one soggy day in college that taught me how the spiritual disciplines of climate action can form our hearts. I had just finished a marathon day of classes on campus and was slowly navigating the byzantine city bus system on my way back home. Underutilized and underfunded, the public transportation looked a lot like that of most midsize US cities: slow, inconvenient, and unreliable.

Yet, I had made a commitment to ride it as often as possible in lieu of driving to and from campus alone in my car. I had just transferred from my first bus and was waiting for the second bus of my journey to carry me home. The bus was late (a common occurrence), and the sky finally made good on the threat it had been making all day and ripped open, drenching me in seconds. I scrambled up the nearest stoop and took shelter beneath the portico of a total stranger, praying the homeowner wouldn't see me and banish me back into the deluge.

Quickly enough, the offending cloud drifted on to visit its displeasure upon some other unlucky souls, and I squished and squelched my way back to the bus stop. Just as my internal pity party was getting into full swing, a dazzling sun broke through the clouds, and I experienced a moment of clarity I'll never forget: I was not the center of the universe. The message drilled into me from birth that I deserved to get wherever I needed to be as quickly as possible, regardless of the climate cost or the damage done to other beings, was wrong. I could wait. Maybe, just maybe, it was good and necessary for me to wait, dripping and uncomfortable, and to remember that every single one of my wants—like getting home as fast as possible and staying dry in the process—need not be met all the time. Maybe I needed that soggy reminder that my human calling to serve and protect creation was a call to responsibility rather than privilege, and that sometimes responsibility costs us something.

As a newfound understanding slaked my soul and the warm sun dried my body, I began to feel joyful at my reorientation to my proper place in the order of things. And joy was to be expected. After all, "joy is the keynote of all the disciplines," says Richard Foster. Their purpose is "liberation from the stifling slavery of self-interest and fear."[10]

Joy, not drudgery, is the promise of the spiritual disciplines of climate action.

God's Pleasure, Our Joy

When it comes to incorporating personal actions into our daily rhythms to better align our lifestyles with the dire reality of the climate crisis, it's important that we resist the urge to ground our action in guilt, shame, and self-punishment. There is certainly an element of sober responsibility. As those creatures who alone have been tasked with the obligation to care for our fellow creatures, it is imperative that we recover a sense of creational accountability for the ways in which we live and move and have our being on the Earth. Yet, this accountability must always be balanced with joy, for when we are living properly into our God-given role as earth keepers, joy abounds. That's just the way God made us. In the film *Chariots of Fire*, Scottish Olympian and Christian missionary Eric Liddell famously says, "God made me fast. And when I run, I feel God's pleasure."[11] In the same way, every human ever born can say, "God made me an earth keeper. And when I serve and protect creation, I feel God's pleasure."

It's likely you know by now those steps you can take to better protect creation in your everyday life. I have a lengthy list that I often bring with me when I give a talk to prime people's imagination. It includes all the usual suspects: eat less meat, fly less often, wash laundry in cold water, hang dry your clothes, take public transportation or ride your bike as often as possible, compost your food scraps, try to keep just one car for your family and make it electric if you can, sow a vegetable garden in your back yard.

Rather than fill these pages with more, there's an appendix in the back with the rest of them. If you want even more solutions, check out Project Drawdown (drawdown.org) and Living the Change (livingthechange .net) for a list of the most effective steps you can take as an individual in your everyday life to impact the climate crisis.

Consider some of these steps that you can take in your daily life, and as you do, perform them as spiritual disciplines. Be meditative and mindful as you practice them. Go about them not as an act of penance but as an act of love for God's creation and freedom from the slavery of consumption and infinite accumulation. Expect joy.[12]

8

LOVING OUR NEIGHBORS
IN PUBLIC

WHEN I SPEAK ABOUT THE NEED FOR CHRISTIANS to cultivate spiritual disciplines of climate action in their personal lives, I'll often receive pushback that sounds something like this: "Spiritual formation is important and all, but we're in a crisis here! We don't have time to fritter around the edges with personal action that will do little to address the climate crisis we face at the speed and scale that justice demands."

While I disagree that personal transformation is inconsequential to addressing climate change, this response is right in one regard: we are facing an imminent crisis. As we saw in the introduction, climate experts are remarkably united in their assessment that climate change is real, it's us, and it's bad. World leaders have not done nearly enough to address this reality with the urgency it requires.

Experts tell us that the world must do all it can to limit warming to 1.5° Celsius above preindustrial levels to avoid severe climate impacts. However, as of 2021, we were on pace to blow past that threshold by 2040 if drastic emissions reductions aren't implemented.[1] To achieve these drastic changes, every nation in the world would need to implement "rapid and far-reaching" transitions in energy, land, infrastructure, and industry on a scale never before seen in human history.[2] Without significant adjustments right away, we will begin to see catastrophic impacts and irreversible tipping points by 2030, even before we cross the 1.5° Celsius Rubicon.[3]

In the face of these long and dire odds, how can washing clothes in cold water and eating less meat avert climate catastrophe? It can't—at least not on its own.

Climate change is a systemic problem that requires systemic solutions. It is the result of a million collective decisions we have made as a species about how to grow and move our food to get it from fields to mouths; how to heat, cool, and light our homes; how to get from point A to point B. For most of human history, how to safeguard human survival, maximize human comfort, and advance human progress were open questions. Various cultures and societies had different answers: hunting and gathering, nomadism, agrarianism.

Periodically, answers to these questions were so compelling and effective that they achieved almost species-wide adoption. About ten thousand years ago, humans discovered that rather than merely supplementing their hunting and foraging with a handful of their favorite cultivated foods, they could instead plant and harvest most of the food they needed to survive. As people stayed put to plant, tend, and cultivate their crops, they began to cooperate in new ways. Societies and cultures began to develop, and primitive cities were born.

In the European late Middle Ages, new ideas about the fundamental nature of the world and human knowledge of it began to emerge. Copernicus, Galileo, Francis Bacon, Isaac Newton, and others challenged the prevailing notion that the world was intelligible only through ritual and mediated primarily through divinely ordained institutions. Instead, they argued, objective reality could be discerned by all through basic scientific methodologies. This shift in how humans came to know and observe what was true found its intellectual climax in the Enlightenment. Through empiricism and methodical study, access to objective truth was democratized. No longer did unapproachable priests or inscrutable rituals hold the keys to transcendent knowledge. Instead, the answers were to be found in human reason, and they were accessible to everybody.

Not long after this epistemological revolution came a technological one. Steam and water power began to be harnessed, new chemical and metallurgical processes were discovered, and mechanized production began to churn out consumable goods on an industrial scale. The Industrial Revolution represented a tectonic shift that would touch every aspect of human life, including how humans would answer the basic questions of survival, comfort, and flourishing for centuries to come.

However, one major question remained: how to power all these burgeoning new industrial techniques. Charcoal and biomass didn't burn hot enough for the more advanced industrial processes, and Europe had already denuded most of its forests anyway. The fateful answer came in the form of the black, sooty lumps that were already widely used in Britain for domestic heating. Now, coal would be enlisted on an industrial scale. It's adoption would spread from Britain to the rest of continental Europe and, due to the colonial reach of Britain and the rest of Europe, would eventually be exported to much of the rest of the world in the late nineteenth and early twentieth centuries.

Now humanity had new answers to the questions about survival, comfort, and progress, and each of them was built on the foundation of cheap fossil fuels. Whereas humans in an earlier time might have answered these questions in a variety of ways and used myriad techniques to do so, by the turn of the twentieth century the hegemonic powers of globalism and colonialism turned Europe's answers into the world's answers.

Save for subsistence farming and economies unable to achieve industrial development, nations would now grow their food in vast monocultures of commodity crops. Thanks to the Green Revolution of the mid-twentieth century, the depleted soils would be reenergized through synthetic, fossil fuel–based fertilizers. The vast acres of crops would be protected from pests and weeds alike through the application of fossil fuel–based herbicides and pesticides, and they would be harvested by enormous internal combustion machines fueled by diesel.

The coal-powered steam engine would connect people like never before. Railways soon crisscrossed entire continents, moving people across previously unfathomable distances in shockingly short order. Horses and buggies would be rendered obsolete for all but the shortest journeys. Even those trips would be transformed by Henry Ford's assembly line and Standard Oil's aggressive exploitation of North American oil fields.

Thomas Edison's innovations to generate and distribute direct-current (DC) electricity from centralized power plants to homes furnished with his patented incandescent lightbulbs in 1882 Manhattan would spark the possibility of a revolution in the way we power our homes. Nikola Tesla, George Westinghouse, and others would then move global electrification from possible to guaranteed with their innovations in alternative-current (AC) generation and transmission.

These systemic answers to the basic questions of survival, comfort, and progress would become further entrenched in global economies, cultures, and societies over the subsequent decades of the twentieth and early twenty-first centuries. A vast web of laws, regulations, incentives, and jurisprudence has made industrial agriculture, internal combustion transportation, fossil fuel–powered electricity generation, and countless other societal practices and norms easier, cheaper, and more convenient. Unfathomable fortunes have been made by those able to extract and market coal, oil, and natural gas. Multinational energy corporations, with an assist from the US Supreme Court's *Citizens United* decision in 2010 granting them unlimited political influence, have come to wield unprecedented power. Today, 71 percent of global greenhouse gases are emitted by only a hundred companies.[4]

The result of all of this is that out-of-control climate change is the sum of countless constitutive parts—parts that have been shaped, guided, and advanced through specific policy decisions that we as a society have made. If that sounds depressingly bleak to you, you're not alone. After all, how can any of us possibly do anything to affect change within such an opaque, labyrinthine system? It's an important question with which to wrestle. It's necessary for us to appreciate the size and scale of the challenge we face.

Yet, even as we respect the scale of the challenge, we must remember that the challenge is one of humanity's own making. There was nothing preordained about the ascendency of fossil fuels. In many ways, it was simply an accident of history. Fossil fuels were there for the taking, so humans took them—and, it must be said, did extraordinary things with them. Fossil fuels have pulled hundreds of millions of people out of poverty. They've led to life-saving advancements in medicine and technology. However, these advancements have come with staggering costs as well—to human health, to biodiversity, and to long-term climate stability.

In the same ways that Alfred Nobel, John D. Rockefeller, Edison, Tesla, and countless others contributed to the establishment of a fossil fuel economy, human ingenuity and innovation can be unleashed to reshape society in ways that are healthy and sustainable. In fact, many of these innovations have already been discovered—from utility-scale solar and wind energy to electric vehicles to exciting advancements in battery technology. They simply need the opportunity to penetrate global markets and achieve wide-scale adoption. To do that, they need the same policy supports that fossil fuels have enjoyed for the last 150 years.

And if that doesn't sound like good enough news, here's the really good news: in liberal democracies like the United States and most other Western, industrialized countries, we are the ones who make policy. Sure, we elect politicians to do the nitty-gritty sausage making, but the buck always stops with us. Elected officials derive their power from us. Policymakers write policy that we, collectively, tell them to. Through our votes and our engagement, through our activism and our pressure, through our voices and our presence, we are the ones who shape laws and policy.

I understand if this might feel like cold comfort to those of us living in what often feels like the death throes of democracy. Authoritarianism is on the rise around the globe, and in the United States we are seeing increasing illiberalism, the shattering of democratic norms, emboldened attempts to restrict voting access, and a government that seems less responsive than ever to the will of the people. And as easy as it can be to allow this to lull us into apathy and cynicism, it is precisely the opposite—action and agency—that will save us.

Injustice thrives on apathy. As long as those of us who care are too disillusioned to resist, it will continue to flourish. After all, as John Stuart Mill quipped in 1867, "Bad [people] need nothing more to compass their ends, than that good [people] should look on and do nothing."[5]

Let us be clear on one thing: the status quo is not neutral. The world in which we live and move and have our being has been shaped by laws and policies that benefit some to the detriment of others. By abstaining from the political process—with all its messy ambiguity, compromise, and disappointment—it can be easy to convince ourselves that we are washing our hands of it. We tell ourselves that there may still be evil in the world, but at least we aren't to blame. This is a lie.

By simply living our lives amid the structures and institutions created by these laws and policies—by buying lettuce from the supermarket, filling up our car at the gas station, or just flipping on a light switch— each of us is complicit in the injustices of the status quo. This means each of us has a moral responsibility to do what we can to create a status quo that is healthier, safer, and more just for everybody.

Silence in the face of injustice speaks volumes. Action does too. Only one question remains: What do you want your life to speak? The Jewish Talmud offers a powerful answer. In its commentary on Micah 6:8, it states, "Do not be daunted by the enormity of the world's grief. Do justly now, love mercy now, walk humbly now. You are not obligated to complete the work, but neither are you free to abandon it."

Acting Like Citizens

So how do we begin to "do justly now" when it comes to the tangled web of special interests, private profits, externalized costs, and policy preferences that are driving the climate crisis? By acting like citizens.

If asked what the primary act of citizenship is, many might say voting. Voting is indeed a critical component of our citizenship. Yet, by itself it is incomplete. It leaves us listless for large swaths of the calendar, enlisted to do our civic duty once every two years (or maybe more often for those ballot box warriors dutifully casting their ballots for every off-year bond millage proposal and school board election) and then left to wait for the next election. It lulls us to sleep to the fullness of our civic life. More insidiously, it provides ample opportunity for special interests to fill the vacuum caused by our civic absence between elections and to ensure that public decisions benefit them rather than the common good.

It also tends to reduce our civic life to what political scientist Eitan Hersh calls "political hobbyism."[6] When our only civic activity occurs once every couple of years, we are left engaging politics and civic life less as a group activity and more as a spectator sport. We consume vast amounts of political media and post angry diatribes on social media, but we don't actually *do* anything. Aristotle once observed that humankind "is by nature a political animal."[7] If this is true, then political hobbyism reduces us to domesticated lap dogs.

To say that voting is all it takes to act like a citizen is like saying eating lots of fruit is all that's needed for a healthy diet. True, fruit is vital for healthy bodies, but without vegetables, protein, and crucial vitamins and minerals found in other food sources, our bodies will soon begin to atrophy. We need a diet filled with varied sources of a diverse range of foods in order to be healthy.

So too with our body politic. We need citizens actively engaging in a wide range of civic activities meant to include as many diverse voices as possible in public conversations, to pressure policymakers to enact laws that advance the common good, and to cast a vision for a common life where everybody can flourish and thrive. And this wide range of civic action has a name: advocacy.

As Christians, we have a particular responsibility toward this kind of engagement in public life. Jesus taught us that the most important things we can do to follow closely after him is to love God completely and to love our neighbors as if the shoes we are standing in belong to them. While the question "Who is my neighbor?" has been asked throughout the centuries, Jesus seems to have answered it definitively in Luke 10:25-37. A teacher of the law asks Jesus what he must do to inherit eternal life. Using the Socratic method, Jesus gets the teacher to answer his own question: love God and love my neighbor. But the teacher of the law isn't satisfied with this answer. After all, who exactly is my neighbor? Must I show loving kindness to all people, or can my compassion be circumscribed in some justifiable way? Jesus responds with the parable of the Good Samaritan, making it perfectly plain that love of neighbor extends further than our cultural sensibilities would like or even imagine, even to our enemies, as Jews and Samaritans were in Jesus' time.

In the hyperconnected, globalized world of the twenty-first century, where the food we eat passes through dozens of hands and traverses multiple continents, where flipping on a light switch using electrons generated by fossil fuel combustion has direct consequences for a Bangladeshi farmer's ability to feed her family, and where driving our cars down the interstate sends toxic fumes across multiple state lines, there can be little doubt who our neighbors are. The only possible answer to the question "Who is my neighbor?" in the twenty-first century is "everyone."

As much as we may want to—as much as the teacher of the law wanted to—we simply do not have the luxury of hiding behind hermeneutical gymnastics to absolve us of our moral responsibility to everyone and everything in God's good creation. Our most basic actions, as people living in the modern world, have direct consequences for people living half a world away. And the reason is because of policy and politics—those society-wide decisions we have made about how to answer the basic human questions of survival, comfort, and progress.

So how do we love our neighbor in a globalized, interconnected world shaped by policy that privileges fossil fuels and elevates some lives while sacrificing others? In a word: *advocacy*. Advocacy is what it looks like to act like citizens and to love our neighbors in public.

Telling your federal lawmakers that you support policies that will end the tax credits and subsidies funneling billions of dollars to the fossil fuel industry every year and will instead create a path toward a 100 percent clean-energy economy is a way to love your neighbor suffering from food insecurity due to climate-induced weather variability. Testifying at a local zoning board meeting against the siting of a toxic coal ash dump near predominantly Black and Brown neighborhoods is a way for you to love your neighbor suffering from the legacy of environmental racism. Writing a letter to the editor or an opinion piece for your local newspaper sharing why your faith compels you to act on climate change is a way to love your neighbor by telling them that you believe them when they say that climate change is killing them, and that you are ready to stand beside them in public solidarity. Marching, protesting, and nonviolently demonstrating against the construction of a pipeline that violates treaty rights and threatens the lives of Indigenous communities is a way to use your body to love your Indigenous neighbors.

Leveraging your social media channels to amplify the stories of climate victims, to highlight potential solutions, to push back on misinformation, and to offer a concrete example of what Christian climate action can look like is a way to love your neighbor too. By doing so you keep the climate crisis unapologetically in front of your followers and build toward that positive social norm where other Christians can also see themselves in the story of action.

Voting for candidates who will implement policies to create millions of healthy, family-sustaining, clean-energy jobs is a way to love your neighbor who will be out of work when the coal mine or the oil field in town inevitably goes under. Supporting candidates who run on a platform of a 100 percent clean-energy economy and advancing environmental justice is a way to love your neighbor who is harmed by climate impacts by showing them that you believe their suffering is not inevitable, that a different world is possible, and that you are ready to dream that world into existence alongside them.

All of it is advocacy. Cornell West said, "Justice is what love looks like in public."[8] Advocacy is *how* we love in public.

Calling Congress

So how do we begin to love our neighbor in public through advocacy? We make a start, knowing that we are not obligated to finish the work, but neither are we free to abandon it.

One of the simplest ways we can start is just by picking up the phone and calling our members of Congress to tell them that, as Christians, we expect them to enact policy that averts the worst consequences of a changing climate. Now, I know that "simple" and "picking up the phone" are oxymorons for a lot of millennials and Generation Z folks who have grown up without a landline in the house and who text much more often than they speak on the phone. As a millennial who'd rather text than call ninety-nine times out of one hundred, I get that phone calls to strangers can be stressful. But I'm here to tell you that calling Congress is so much easier than you think it is, and it matters much more than you might know.

For instance, did you know that the US Capitol has a switchboard number? It's 202-224-3121. When you call it during business hours, a human usually answers. You simply let them know the name of your senator or representative. (Don't know who they are? Google it before you call.) Once you are connected to the right office, another very friendly human answers. This is usually a young (early twenties) and aggressively underpaid intern. The interns are almost always unfailingly kind. You can simply tell them that you'd like to leave a message for the senator or representative, and they will invite you to go ahead when you're ready. Then you share just two to four sentences' worth of input. It could be to urge support for a particular bill, but it could also simply be your own story about why you, as a Christian, are concerned about climate change and why you want to see your senator/representative do all they can to address it.

If you've never called your member of Congress and are feeling a little nervous, go ahead and call after business hours. You'll get an automated message that asks you to enter your zip code, and it finds your senator or representative for you. It'll connect you to the office voice mail for you to leave a message. You can even write out what you want to say ahead of time and just read it off!

Or better yet, you can use this script:

> Hi, my name is (name) and I'm a constituent from (city). I don't need a callback. I'm calling because, as a Christian, I believe that loving God and loving our neighbors means taking climate change seriously. I would like to see the representative/senator show leadership on climate change by supporting policies that move us toward a 100 percent clean, renewable energy future. Thank you very much.

A lot of people think they need to be policy experts before they do this, but you don't! You don't need to know bill numbers or a bunch of stats and figures. You're not going to be quizzed. You simply need to share three things: (1) you are a constituent, (2) you are a Christian, and (3) because of your faith, you are passionate about addressing the climate crisis and want to see the senator/representative do all he or she can to address it. Or, to put it in the language we've already used, you don't need to know policy to make a difference. You just need to know your own story.

I've trained hundreds of people in this kind of congressional advocacy. Almost to a person, I hear anxiety and apprehension before they try it. And almost to a person, I hear surprise after it's over—surprise at how simple it was, how casual the conversation was, how empowering it felt. I get that it can feel intimidating to consider calling the office of your member of Congress if you've never done it before. It was for me too. But remember: they work for you, not the other way around. This is the way our political system is supposed to work.

Maybe it isn't intimidation you feel about calling your members of Congress but cynicism. After all, how much difference can a phone call make? And how can we be sure our messages even get to the senator or representative? What happens to our comment after we hang up with the intern? Does it go into a black hole, never to be seen again?

I actually asked this question of a staffer at one of my first in-person meetings with my senator's office years ago. After a very cordial meeting with the senator's legislative aide, we spilled out of the conference room and were milling about near the reception desk. We were exchanging pleasantries before we took our leave when I blurted out the question that was burning inside me for most of the meeting: "So what happens next? Will the senator even know we were here?"

The staffer looked surprised for a moment before his face broke into a smile. "Of course!" he said. Then he shared with our group their process for communicating constituent concerns to the senator. Every single meeting, phone call, email, piece of mail, or other correspondence was meticulously logged in a spreadsheet by issue and was assigned a number value based on the amount of effort it took the constituent to send it. A handwritten letter was assigned a higher value than an email because the extra time it takes to write, address, stamp, and mail a letter communicates a higher level of dedication. Our in-person meeting, he told us, was assigned one of the highest number values. At the end of each week, they have an all-staff meeting with the senator where they share the numbers with her for each issue over the previous week, as well as some major themes they heard from the communications. If a number for a particular issue is high over several consecutive weeks, that signals to the senator that it is something her constituents are particularly concerned about. If it remains high moving forward, she directs her staff to dedicate more of their time to it. In other words, our communications to our members of Congress are a direct signal to them about where they should be directing their legislative time, attention, and resources.

I was stunned to hear there was such an organized system operating behind the scenes of our meeting. "Is this unique to your office, or do other offices have a system like it too?" I asked. "Every single congressional office does the same thing," he replied with a smile.

I know it can feel fruitless sometimes to pick up the phone and leave a message with an intern, hoping that it will somehow reach the boss.

But I'm here to tell you that it isn't. And if everyone picks up the phone, sends an email, writes a letter, tags a congressional office on social media, or holds an in-person meeting about climate change action—and keeps doing it—it will change everything. I guarantee it.

So why not give it a try? Grab your phone right now and call that switchboard (202-224-3121). I'll wait. And when you're done, save it in your phone as a contact so that you can keep picking up the phone and do it again and again until those weekly staff meetings become climate-policy strategy sessions for every single office on Capitol Hill.

Writing for Change

Another crucial act of public neighbor-love is going on record through public writing. Though the influence of the opinion pages of your local newspaper has waned in recent years, reader-submitted articles are read religiously by congressional offices. What's more, they remain some of the most read pieces published by print and online outlets, period. Simply put, people are interested in what their neighbors have to say.

If you've never submitted an opinion piece (reader-submitted columns often called "op-eds" because they historically ran opposite the editorial page in print newspapers) or a letter to the editor (LTE) before, I understand that it can feel like a total black box. You may feel that you don't have anything to say, or that you'll never be able to get a piece placed even if you do figure out how to write it. The truth, though, is that op-eds and LTEs are much easier to write than you may think and that unless you're pitching to the *New York Times*, most editors are running skeleton newsrooms and are desperate for compelling, high-quality, reader-submitted content.

First, though, let's understand the difference between an op-ed and an LTE. A letter to the editor is a short piece (think 150 to 250 words) that is either about a current event or a response to a piece recently published by the outlet. These are those blurbs you'll see at the front of magazines like *The Atlantic* or *Time* reacting to articles that ran in the previous month's issue. Whether reacting to a previously published article or sharing a viewpoint about a current event, the successful LTE writer will choose one point to make and will seek to make it as persuasively and succinctly as possible. The shorter format makes them a little easier to craft, and they are typically relatively easy to get placed.

An op-ed, on the other hand, is a longer-form, argument-driven piece (think 650 to 800 words) that is less timebound than an LTE and makes an argument that will still be relevant weeks, months, or even years after publication. As with any genre of writing, it has its own structure that editors look for when reviewing submissions.

The best op-eds are structured thoughtfully, with a catchy lede (the piece's first one to three sentences), a thesis, two to three pieces of thesis-supporting evidence, a couple of anticipated counterarguments with thoughtful rebuttals (called "to be sures"), and a succinct conclusion. To save room here, I've included more in-depth details on how to craft successful LTEs and op-eds in appendix B.

This format is a framework, not a formula. In most op-eds, these elements are woven throughout, with evidence-based arguments sneaking into ledes, a thesis woven throughout the entire piece, and "to be sures" peppered everywhere along the way. The most important takeaway is that successful op-eds will contain most of these elements, but the order in which they appear—or whether every element is included— makes very little difference. The most important thing is to write your heart into whatever you craft. If you do that, an editor will sit up and take notice.

Now, about those editors. Even if you craft the most compelling, winsome, inspiring piece a pen has ever put to paper (or fingers have ever typed on a computer), it won't matter unless an overworked and time-crunched editor agrees to publish it. So, how to go about pitching your op-eds and LTEs so that an editor will respond positively?

Pitching an op-ed or LTE is as much a skill as writing one. Like op-eds and LTEs, good pitches follow a structure. The best way to pitch is via email, and your email should do several things, all in very few words. Remember, editors are busy, and they get a lot of emails! They need to know, within the first few sentences, whether it's worth their time to read the draft you are sending them.

To help them do this, your email pitches should answer three basic questions in 150 words or fewer: Why me (why are you the right person to write this piece)? Why now (what's the connection to current events)? And so what (why will the editor's readers care)?

Easy, right? Hardly! In my experience, crafting a 150-word pitch email is often harder than writing the 800-word op-ed I'm trying to get an editor to read. But if you answer these three questions, include your thesis in the first couple of sentences, and copy/paste your piece below in the email body, you'll give yourself a great chance. If you're not sure how to find the contact info for an outlet's opinion editor, the Op-Ed Project has you covered. Head to theopedproject.org/submission -information to find editor contact information for hundreds of newspapers across the United States. And check out appendix B for more detailed advice on writing a successful pitch.

Finally, expect rejection. Rejection is a feature of this process, not a bug. I have had pieces rejected seven, eight, nine times before getting an editor to say yes. I have pieces that I thought were terrific that endured a parade of rejections and have never seen the light of day. There are a whole bunch of reasons why an editor turns down a pitched piece. Learn from their feedback, dust yourself off, and keep pitching!

Taking to the Streets

You may have noticed an uptick in protests, marches, and demonstrations over the last several years: the student-led March for Our Lives against gun violence in schools, the #FridaysforFuture classroom walkouts and marches for climate action organized by Greta Thunberg, the Women's March, the March for Science, the massive racial justice protests in the summer of 2020 in response to the murder of George Floyd. There is no doubt we are experiencing a political moment of mass mobilization.

On the one hand, this is a sign of a vibrant civil society engaged in public decisions about how to advance the common good. On the other hand, few of these mass demonstrations have led to the kind of meaningful policy change that they are seeking—at least, not yet. Gun-control laws remain largely unchanged. Clean-energy and climate policies are beginning to break through and become law—especially at the state level, and most notably with the passage of the federal Bipartisan Infrastructure Investment and Jobs Act in 2021 and the Inflation Reduction Act in 2022, which together invested over 400 billion dollars into clean energy and transportation. Yet much more is needed to achieve change at the speed and scale demanded by Greta and her cohort. Juneteenth was declared a national holiday, but the far more consequential police reforms that the Movement for Black Lives sought fell apart in 2021 at the eleventh hour.

It's true that it takes time for mass mobilization to translate into policy change. Yet, the question must also be asked: Why are so many of these major movements failing to translate the energy and passion of millions of people into meaningful, lasting change?

Participating in marches and demonstrations tends to encourage what we might call "advocacy tourism." People come out for a one-off event and feel the high of experiencing collective action, then return to their regularly scheduled lives. No follow-up calls to Congress are dialed. Few op-eds are penned explaining why we were in the streets at all. A kind of moral licensing can occur that makes us feel good about being active ("I did my part!") while excusing ourselves from the rest of the responsibilities of civic life.

We saw a lot of this kind of advocacy tourism with the racial justice protests in 2020. Well-meaning people were rightly shocked and horrified by the murders of George Floyd and Breonna Taylor and the larger racial inequities that they revealed in our policing and criminal justice systems. They channeled their grief and anger into some of the largest multiracial demonstrations the country has ever seen. Some polls estimate that anywhere from fifteen million to twenty-six million Americans participated in a Black Lives Matter protest that summer.

And yet, more than two years after these historic demonstrations, the central goal of the Black Lives Matter movement—meaningful police reform—remains elusive. I can't help but wonder how many of those millions of people merely participated in advocacy tourism—changing their profile picture and marching once or twice—and forgot to do the rest of the work required of them in order to achieve change.

There can be no doubt that protests, strikes, marches, and demonstrations can be powerful expressions of advocacy and potent tools to advance the common good. The modern labor movement of the nineteenth and twentieth centuries, the women's suffrage movement from the Seneca Falls convention in 1848 to the ratification of the nineteenth amendment in 1920, the civil rights movement, the anti–Vietnam War movement—all relied heavily, and to great effect, on mass mobilization to put pressure on the status quo and to create an opening for an alternative to take root and grow. Policy analyst Joseph P. Overton recognized that policies become politically acceptable only when they reach a critical mass of public support. Only when policies become acceptable to the majority of an electorate do politicians feel safe enough to work toward turning them into law.[9]

The Overton Window, as this phenomenon has come to be called, shifts as ideas move from the margins to the center of public discourse. A tried-and-true tactic for shifting the Overton Window of policy possibility is sustained, large-scale mass mobilization in the form of protests, strikes, marches, and demonstrations. The images of Bull Connor's dogs and fire hoses trained on peaceful marches crossing the Edmund Pettus Bridge galvanized public support for civil rights. The Vietnam War was hugely popular until students began demanding transparency and accountability from the US government through widespread, sustained demonstrations.

Yet, by themselves these mass movements did not achieve policy change. They merely created the conditions in which policy change could occur. It took Martin Luther King Jr. in the ear of Lyndon B. Johnson, and millions of engaged citizens bombarding members of Congress, to seize the opportunity created by their marches to get the Civil Rights Act and the Voting Rights Act passed in 1964 and 1965, respectively. It took disciplined organizing and advocacy for almost a hundred years of women's suffrage marches to finally translate into the ratification of the nineteenth amendment.

In other words, to advance our neighbor's good in a warming world, protests, marches, and mass demonstrations are necessary but insufficient. They must be married to other advocacy strategies—like legislative engagement and voting—to ensure that when mass mobilization opens the Overton Window, good policy can jump through it.

So, by all means, march and protest! It is necessary and effective—not to mention cathartic and therapeutic. But we must all be wary of the advocacy-tourism trap and ensure that we are loving our neighbors in public through other advocacy strategies at the same time.

Social Media and Digital Activism

I can almost hear the skeptical groans as I write this. *Social media? Really? That cesspool of negativity, hate, and disinformation? How naive is this guy?*

I get it. Social media are toxic in a million different ways. And yet, I think that social media can provide outlets for productive expressions of modern advocacy for a few reasons.

The first is that congressional staffers monitor social media, and they pay attention to what is being said on their bosses' accounts. Constituents have particular clout, and their comments will certainly be noted.

But how can staffers tell who is a constituent and who isn't? It's not always possible on all platforms, but if you still use Facebook, the Town Hall function allows you to link your address to your account and be presented with a whole host of pages from your local, state, and federal elected officials. You are then able to turn on your Constituent Badge, which shows up whenever you comment on a post from one of your elected officials to signal that your comment is coming from a constituent and not just a troll with too much time on their hands. The icon looks like a little three-columned Parthenon next to your name.

Turn on your badge and give it a try. Comment on a recent post from one of your elected officials, and see if you notice the icon appear next to your name. If you can see the icon, your elected official's staffer monitoring the account can too! If you're using other social media platforms that don't have this kind of function, simply claiming your identity as a constituent in your post or comment is helpful.

A second reason social media should be included in any balanced advocacy diet is because you have countless eyes on you when you use social media. You have no idea who might read your post or be encouraged by a comment you make. I have heard from a lot of people—some of whom I didn't expect—that they are encouraged by the way I use social media, especially with how I relate to less-than-civil commenters. I'd be willing to bet that you have people in your social media ecosystem watching you and being encouraged by you too.

We all know social media platforms are some of the worst places to have a debate. It turns out, though, that they can be pretty terrific places to encourage and inspire. That's why, for the most part, I think it's worthwhile to engage with negative comments that we may receive on our posts. (Disclaimer: Only you can determine whether this exercise would be healthy or harmful for you. Take care of yourself and your mental health. If that means leaving a negative comment alone, do it.) We don't engage for the sake of the person making the negative comment. We're not likely to move them. Instead, we can respond for the sake of the people watching the exchange. We can model respectful disagreement, principled pushback against misinformation, and fact-based persuasion. There are more people than you likely know watching you on social media, so why not use your influence to encourage, inspire, and inform?

Remember George Marshall back in chapter six, talking about creating that positive social norm for action by modeling for others who share your identity and values that people just like them can and are involved in climate action? Harnessing the amplifying power of social media is a powerful tool for creating a positive social norm among our fellow Christians that climate action is perfectly consistent with who they are and that a whole lot of other brothers and sisters are already acting.

A final reason social media can be a worthy tool for effective advocacy is because of the power that can be built and unleashed through it. It's entirely possible most of us never would have learned the name George Floyd if a teenager named Darnella Frazier hadn't pulled out her phone to record his murder and then post it to social media, where it was shared around the world, sparking outrage and action. The phrase "Black lives matter" would not be in our national lexicon if three friends hadn't been grieving together on Facebook the acquittal of George Zimmerman, the man who shot and killed Trayvon Martin, when one of them wrote the epitaph out of her pain and frustration. Many of the Arab Spring uprisings from 2010 to 2012 were, at least in some part, facilitated through activists' use of social media to organize and promote protests on the ground. Even fans of celebrity artists, like Taylor Swift's Swifties and the BTS Army (so named after the K-pop boy band BTS), have coalesced on Twitter, TikTok, Reddit, and other platforms to engage in digital activism ranging from overwhelming hateful hashtags with positive messages to overloading an app being used by the Dallas police department to identify Black Lives Matter protestors in 2020.

Social media have enormous power. Most of the headlines are dominated by the times that these sites get used for ill, but what if we started harnessing them for good? What if we used social media to communicate directly to our elected officials? What if we engaged these tools to model constructive dialogue and signal to our networks that we care enough about climate change to speak out? What if we used them to organize and activate? It doesn't matter how many followers, friends, or subscribers you may have. All of us have a sphere of influence. All of us can use social media to make a difference.

Voting

Of course, we couldn't talk about loving our neighbors in public in a warming world without talking about voting. In a democracy, voting is one of the most concrete expressions of institutional neighbor-love there is. It is an expression of hope in our common life together. Even when our public circumstances are bleak, we can remain committed to each other at least enough to keep engaging in the process of participatory decision making. Whenever I go into a voting booth, I can't help but think of Jeremiah, who, after declaring the impending collapse of the kingdom of Judah at the hands of the Babylonians, went out and bought a field there anyway and proclaimed that "houses, fields and vineyards will again be bought in this land" (Jeremiah 32:15).

We hear a lot about how broken and imperfect our electoral system is: rampant gerrymandering, rafts of dark money flowing in all directions, corrupt public officials, restrictive voting laws designed to make it disproportionately harder for some people to vote than for others— the list could go on. These and others are often cited as the reasons for people opting out of the electoral process altogether. When I hear this, though, I can't help but wonder: What did you expect? After all, institutions are created and perpetuated by humans, and humans are imperfect. To call a system that's been created by humans imperfect is redundant. It goes without saying.

Some might find this depressing, and I understand why. Yet, in the same way that I find hope in the human roots of the climate crisis, I find hope in the human-ness of an imperfect electoral system because it means that we have the power to make it better. However, we can only make it better if we have skin in the game, if we participate. We can only express our power if we vote, and then do all the other hard and necessary work of advocacy to ensure that the imperfect system keeps getting a little bit better.

I understand how profoundly naive, and more than a little privileged, this may sound. *Easy for the White, male, natural-born citizen to say. He's never had to forgo an entire day's pay or risk being fired for waiting all day in line to cast his vote. He's never had to take three buses and walk half a mile to his polling site.*

Absolutely right. I have not experienced systemic barriers to voting in the same way that millions of my fellow citizens have. That's why I and others like me have a particular responsibility to make sure that we do vote, and then do everything in our power to fight for a system in which everyone has equal access to the ballot. A healthy democracy—where the power resides with the people rather than corporations and special interests—is the only kind of democracy that can deliver transformative climate policy and advance the common good.

When it comes to fighting climate change through the electoral process, researchers are beginning to quantify just how effective our votes can be. The concept of "political carbon offsetting" was coined in a 2021 paper that sought to quantify the carbon-reduction effects of targeted political contributions and of winning votes cast for parties with clear climate action plans versus parties with platforms that ignore or minimize the climate crisis.

The authors looked at the 2019 Canadian election between the Conservative Party of Canada (which wanted to repeal the nation's carbon tax, among other policies opposed by climate experts) and the winning coalition of the Liberal and three other parties that promoted emission-cutting policies. By estimating the emissions reductions that the winning coalition's policies might achieve and distributing these reductions across all voters, the average reduction per voter was 6.7 tons of CO_2e (carbon dioxide equivalent). When they distributed the reductions to only those voters who voted for winning candidates from this climate-action-oriented coalition, the carbon reduction per vote was 34.2 tons.[10]

For scale, going car-free for a year will net you about 2.4 tons CO_2e. Eating a mostly plant-based diet will avert an additional 0.8 CO_2e. In other words, the single most impactful action you can take as an individual to advance your neighbors' good in a warming world is to vote for winning candidates with strong climate platforms.

Of course, there is no one single Christian way to vote. There are a lot of people and groups out there that would like to convince you that there is—that voting for one candidate is the "Christian choice" and voting for others is not. Most of the time, these people are interested less in encouraging Christians toward faithful civic engagement and more in filling their own coffers, burnishing their prestige, and increasing their personal proximity to power. Be wary of Christians who tell you there is only one Christian candidate or only one Christian way to vote.

However, the difficult truth remains that until very recently, only one major political party in the United States has advanced meaningful climate legislation commensurate with the crisis we face while the other has, at best, merely frittered around the edges of lukewarm climate solutions. This means that a balanced advocacy diet is more important than ever for those of us living and voting in the United States. We must be the ones to make climate action mainstream across all political persuasions in our public life. Through raising our voices with our elected officials, writing op-eds and LTEs, protesting, marching, witnessing on our social media channels, voting, and more, we must send a message too irresistible to ignore to all our leaders that strong climate action is the price of entry for public leadership, regardless of party, full stop. Candidates from all parties who are unserious about addressing climate change need not apply.

Advocacy: Our Christian Heritage

These have been just a handful of suggestions for a more balanced diet of faithful advocacy. There are more: community organizing, offering public comment at local hearings, volunteering to be a poll worker, running for office yourself. The point is not to exhaust yourself by doing everything but to find the actions that you enjoy and that are effective, and to practice them regularly.

Now maybe you've read your way through this chapter and have concluded that I'm merely a radical environmental activist wolf in

Christian sheep's clothing. Many of us who grew up in evangelical Christian communities heard little or nothing about practical strategies for political advocacy, and even less about how advocacy is a natural expression of our faith. The thing is, though, faithful Christian advocacy for justice and the common good is not a modern, secular distortion of the one true gospel. It is our Christian heritage.

The gospel has always been political—and politically subversive, at that. Conventional political thinking says that power consolidates at the top among a small group of elites and is exercised on behalf of everyone else, but the gospel says to give power away and let it pool at the bottom of society among the poor and the outcasts. If resources are extracted from the margins of political power and flow toward its center, as they were in Jesus' time and still are today, the gospel says to sell all you have and give to the poor (Matthew 19:21). Jesus was executed by the state, with the support of the religious elites, because his message was viewed as too great a threat to the political status quo.

The first communities that coalesced around the Jesus Way shared food and other possessions, and sold their property and gave the proceeds to anyone who had need (Acts 2:44-45)—a powerful counternarrative to the prevailing political and economic norms of the time. Many of the first Christian martyrs were killed for refusing to acknowledge the deity of the Roman emperor, a profound act of political protest. Christians from William Wilberforce to Cesar Chavez to Martin Luther King Jr. understood advocacy as an act of neighbor-love and engaged in advocacy to advance God's justice for their enslaved neighbors, their migrant worker neighbors, and their Black and Brown neighbors.

There are some who will say that political advocacy is too controversial. They fear that advocating for specific issues Scripture clearly identifies as close to the heart of God—like a healthy and thriving creation where all people can experience abundant life—is too divisive. They don't like politics because it's too charged, too messy.

While the impulse behind this thinking is understandable, I simply can't agree with it. As an act of institutional neighbor-love that is firmly rooted in our Christian heritage and that takes its lead from Scripture, political advocacy is not controversial or divisive—it's faithful. It's an act of discipleship. It's when we stay silent about the things Scripture shouts that we are being controversial. It's when we refuse to love our neighbors through advocacy that we are being divisive. This is all the more true in a rapidly warming world.

If, as we saw in chapter four, the gospel is good news for all creation, especially for the poor, and it is advanced by the Holy Spirit working through us, then our joyful response to the gospel can't afford to opt out of concrete opportunities to partner with the Spirit to enact policies that are good for creation and for the poor. Scripture is not neutral on matters that matter to God—justice, peace, wholeness, dignity for created things. Why, then, would we believe that we can somehow be neutral in a broken and sinful world and still preach a compelling, authentic gospel? Why would we think for a second that we could somehow remain above the fray by opting out of advocacy, civic engagement, and politics?

The truth is, most people aren't actually turned off by politics. When people say they hate politics, most of the time what they mean is that they hate partisanship. To be fair, the two have become synonymous in our hyperpolarized public square, but politics doesn't have to be partisan. It doesn't have to be craven and cynical. And we, all of us, get to choose how we will participate in the public square for the sake of climate action and the common good.

However, if our advocacy isn't driven by partisanship, what is it driven by? Is there a vision for faithful Christian citizenship that can sustain our civic action in ways that love our neighbor and honor God in a warming world?

9

CHRISTIAN CITIZENSHIP IN A WARMING WORLD

THERE'S A STORY THAT FLOATS around faith-based advocacy circles that goes something like this: There was once an elderly priest who had made it a habit in his later years to become well known around the office of his congressional representative. He would pop in seemingly all the time. New staff would often be taken with his mild manner and clerical collar. And while he was exceedingly kind, more senior staff knew better. They knew that the priest came to their office not to chat or to offer pastoral care, but for business. He came to prophecy, to admonish. He came to advocate.

Nobody in the office understood this better than the representative. Even though staff were often on the receiving end of his jeremiads, he reserved his most pointed rebukes for the representative herself, whom he insisted on speaking to directly every time she was in the office. His talking points rarely deviated, bathed in scriptural imperatives and always focusing on defending the dignity and well-being of society's most vulnerable members.

After a particularly extended visit from the priest, the representative finally let her frustration get the better of her. Cheeks flushed, she blurted out, "You keep coming to my office and telling me all the ways that I'm failing. You point out all the ways that current policy is leaving vulnerable people behind, but you never tell me what you want me to do about it. So, what? What do you want me to actually do?!"

The kind smile never faltered from the priest's face as he received the representative's outburst. Gently, the priest replied, "Congresswoman, it is my job to demand that justice roll on like a mighty water. It is your job to figure out the plumbing."

Christian Citizenship in a Warming World Takes Its Cues from Scripture

This modern-day parable may be a bit tongue-in-cheek (I do think we need to bring some solutions to the table!). However, I believe it offers us several insights into how faithful Christian citizenship in a warming world can serve as a counterweight to our current hyperpolarized, unproductive civic life.

First, Christian citizenship in a warming world takes its cues from Scripture. While it is tempting in today's toxic public square to base our advocacy on the talking points from partisan politicians, the latest breaking news headline on our favorite twenty-four-hour news channel, or the social media groupthink from the various echo chambers cluttering our feeds, it's necessary for us to remember that as Christians we engage in public life differently. Our public priorities must be shaped by the concerns of Scripture and by the concerns that Jesus focused on in his ministry. After all, Jesus is the one, true Word (*logos*) of God and the interpretive key for the rest of what Scripture has to teach us.

When it comes to advocacy for climate action, we can be assured that our efforts rest on solid scriptural ground. As we saw in chapter three, all of God's creation is a main character in the Big Story of God's saving work in the world. God calls creation good, and he tasks his image bearers to work for its full flourishing through service and protection. God is in intimate, immanent relationship with all of creation, and his heart breaks at the totalizing effects of sin in the world. So God takes on the stuff of creation himself in order to rescue all of it from death-dealing distortion and, ultimately, to bring all of it back into restored right relationship with him once more.

In chapter four, we explored how the gospel itself is not merely good news for human souls but for all creation. We noted that Jesus himself declares that the fundamental shape of his rescue mission on earth is to proclaim this good news especially to the poor, the prisoner, the blind, and the oppressed (Luke 4:18). Later in his ministry, Jesus teaches that our response to the Big Story of God's saving work in the world could be summed up in two commands: love God with everything you've got, and love your neighbors as if their current circumstances and future prospects were your own.

As Christians in a warming world, then, our climate advocacy is no mere passing fad or act of performative wokeness. It is a direct response to Scripture's own definition of the gospel: good news for a suffering creation, human and otherwise. It seeks to make that good news known and felt through policies whose results are good for people who suffer most under climate change's unequally distributed weight. Policies that support farmers like Margaret who are being thrown into food insecurity by the millions due to more unpredictable rainfall. Policies that consider homeowners like Robert who have watched their homes—and sometimes their families—be washed away beneath the surging waves or burned to cinders at their feet. Policies that decrease the number of Black and Brown kids inhaling poison from the towering smokestacks in their backyards, suffering asthma attacks and laboring for another wheezy breath, terrified that each ragged one might be their last. And policies that safeguard the prospects of future generations yet to be born, set to inherit a more precarious and desperate world by no fault of their own.

Scripture communicates a multitude of priorities, and these priorities can be faithfully translated into myriad different policy preferences and political expressions. However, Scripture is also clear that any attempts to faithfully translate these priorities into political action must include efforts to address the suffering being visited on people and planet alike by climate change. In other words, when Christian citizenship in a warming world takes its cues from Scripture, climate action is a requirement. The good news of the gospel will simply allow nothing less.

Christian Citizenship in a Warming World Is Other Oriented

If politics is fundamentally about the allocation of resources and the exercise of power, then human nature dictates that it will be inherently selfish. The basic sin lodged deep in every human heart is a self-centeredness that believes that each of us alone knows best, each of us alone deserves most. The rupture reverberating all the way back to the Garden of Eden is a profound lack of imaginative empathy that enables the separation of God, humans, and creation in the first place, and then convinces us that this separation is normal.

The realities wrought by climate change and our collective inability to muster the political will to do much about them are, at root, the fruit of this same basic selfishness. We see it in the rapacious greed of fossil fuel corporations, the cravenness of politicians to fundraise using climate disinformation, and the apathy of millions of citizens who stay silent as millions of our neighbors suffer and die.

That's why the way of the gospel is so radically counterintuitive. The instructions to love our enemies and to pray for those who persecute us (Luke 6:26-28), to turn the other cheek (Matthew 5:39), to love others in the same way we love ourselves (Luke 10:27) only make sense through the lens of the kingdom of God. In God's economy of salvation, self-emptying love on behalf of another is the fundamental shape of reality—be it in the form of a baby in a manger, a bloody cross, or an empty tomb. After all, "the message of the cross is foolishness to those who are perishing, but to us who are being saved it is the power of God" (1 Corinthians 1:18).

If we claim to be transformed by the saving power of that baby, cross, and tomb, then self-emptying love is the shape of our reality too—our entire reality, politics included. For so many Christians, though, our political calculations are still fundamentally selfish. *Our* pocketbooks, *our* safety, *our* social status, and *our* power order our political steps. This certainly seems to be the case for the large majority of evangelicals who voted for Donald Trump in 2016 and 2020, and who continue to support him. Trump has made no effort to disguise his transactional appeals to evangelicals with the language of faith but has instead leaned into the selfishness of human politics by claiming multiple times since 2018 that "no one has done more for Christians or evangelicals . . . than I have."[1] On a stage at Dordt University, a school affiliated with my own denomination, he said flat out that if he were to be elected, "Christianity will have power."[2]

His reward? The overwhelming majority of votes cast in his favor from White evangelicals in both the 2016 and 2020 presidential elections. To all the world, it appeared as though White evangelicals' own safety, security, and comfort were their civic North Star.

A 2020 study from Pew Research illustrates that these appeals to selfishness may have been highly motivating for many of the White evangelicals who cast their ballots for Trump. It found that 67 percent of evangelicals said it is "very important" to them that the president stand up for people with their religious beliefs.[3] In other words, it was very important to a majority of evangelicals that they themselves be defended rather, it would seem, than the poor, imprisoned, and oppressed as Jesus himself instructs (Luke 4:18-19; Matthew 25:37-49).

To be fair, this kind of self-preservation is perfectly human. It is not, however, Christian. It is instead the culmination of decades of Christian citizenship focused on self rather than on others. This civic posture has all too often been motivated by fear rather than love. It represents a public engagement that casts the public square as a means toward maintaining cultural influence and power rather than a means of loving our vulnerable neighbors. Perhaps it's no surprise, then, that a 2019 Barna study found that *evangelical* has come to be viewed by 67 percent of the American population as more of a political than a religious identity.[4]

What if instead of fear-induced selfishness we allowed the other-oriented way of Jesus to guide our civic action? Rather than fear, our public action on climate change would be rooted in a spirit of power and love (2 Timothy 1:7). Rather than letting our politics become all about us, we would instead model the other-focused service of Christ. Rather than privatizing the gospel by limiting its scope only to our individual hearts, we would set its power free to transform the entire world that

God so loves (John 3:16). Rather than cheapening the gospel by trotting it out for cynical photo ops or as a partisan cudgel, we would amplify the good news of the gospel in public by taking actions that are *actually good news* for the poor, the unborn, the marginalized, and all those suffering the effects of a changing climate.

Our first question when considering a proposed climate policy would not be, "What will this policy do for me?" Instead, we would ask, "What would this policy mean for others' ability to flourish and thrive?" Our overriding concern when considering a candidate would not be, "What will their proposed climate platform do for me?" Instead, we would ask, "What will their climate polices do for the forgotten and the dispossessed?" The central goal of our political participation in a warming world would not be self-preservation but the good of those who are harmed most by a changing climate.

The good news is that when it comes to climate policy, this is relatively easy because what is good for others is also good for us. We all want to breathe clean air; drink clean water; have stable, family-sustaining jobs; and live and raise our kids in a world with a healthy and stable climate. There may have been a time thirty years ago when policy tradeoffs had to be made between a healthy environment and a vibrant economy. Today, advances in technology and the growing economic costs of climate inaction mean that policies that lead to a healthier environment also lead to a healthier economy for all.

Christian Citizenship in a Warming World Exudes the Fruit of the Spirit

Maybe one of the reasons that policy solutions to climate change have eluded us for so long is because our politics so often seem to reward the wrong kind of behavior. Transformative climate policies that will rapidly draw down emissions, achieve economy-wide electrification of everything from our cars to our stoves, and ensure that every other nation in the world can do the same will require maturity, compromise, and patient coalition building from our political leaders. Yet, our current politics reward the opposite behaviors. Denigrating a political opponent is rewarded as "being tough." Pandering to the most extreme elements of a party's base with lies and half-truths wins primaries. Outrageous and offensive rhetoric grabs the headlines. And trafficking in conspiracy theories and the demonization of fellow citizens sets fundraising records.

It should come as no surprise, for instance, that by February 2016 Donald Trump had received over $2 billion in free media coverage, dwarfing the earned media of any other presidential contender.[5] That the most outrageous candidate in living memory would also receive the most media coverage makes perfect sense in a system that rewards outrage and grievance.

What a far cry from the values that Paul tells us in Galatians 5:22-23 are the true marks of the Spirit of God in us. It sounds hopelessly naive to believe that values like love, joy, and peace could ever get any purchase in a public square saturated in hate, grievance, and ideological warfare.

Still, I can't help but worry that we write off the fruit of the Spirit in our civic life at our own peril—especially when it comes to the climate crisis. Public conversations about climate change so often feel stuck in what seems like an endless loop of contempt, misunderstanding, and accusation. We are in desperate need of empathy, mutual understanding, humility, and grace if we are to make any progress. These are gifts that Christians are uniquely equipped to offer the public conversation around climate change, but so often we would rather traffic in the political tools favored by the rest of society. By doing so, we allow the terms of our civic engagement to be set for us, and we sacrifice the unique gift that the gospel brings to the public square. Without the fruit of the Spirit, Christians simply become one more special-interest group scraping and clawing for power.

Yet, when the fruit of the Spirit is allowed to shape our Christian citizenship in a warming world, hatred for those with views different from ours is replaced by love grounded in empathy and genuine curiosity about what experiences and values inform their positions. The dangerous soul fatigue that is worn by so many as a perverse badge of honor signaling their devotion to the cause of climate justice is replaced by a winsome joy and profound peace that are rooted in gospel hope and the freedom of knowing that the task of irradicating climate change rests not on our shoulders but on Christ's alone. Scorched-earth ruthlessness gives way to unfailing kindness and unimpeachable goodness.

I've seen what this looks like in the real world, and it's profoundly compelling. In the spring of 2017, I led a group of Christian college students to Washington, DC. They'd traveled on a shoestring from all across the Northeast and Midwest and had already marched in the streets in a peaceful demonstration for climate action two days before. Now, they found themselves in the vaunted hallways of the US Capitol. They spoke in hushed voices, recognizing implicitly the solemnity communicated by the marble columns and soaring archways that ushered them into the chambers of then Senate majority leader, Mitch McConnell. They fanned out, finding space among the richly upholstered furniture adorning the expansive room. They were quiet, but not out of timidity or fear. Their resolute faces and clear eyes communicated resolve. Like the priest in our story, they had come with a message, and they intended to deliver it.

Senator McConnell's chief of staff entered the room and greeted the group. Each in turn introduced themselves. After the pleasantries had been observed, the meeting began in earnest. The students shared passionately about the fundamental injustice of climate change and our Christian responsibility to respond out of love for God and neighbor.

The chief of staff listened politely. When we had finished, he posed a question: "How many of you are conservatives?" The students' hands remained by their sides, and a smirk played across the staffer's mouth as his eyes swept the room. I knew exactly what he was thinking: here was a group of self-styled evangelicals claiming to speak for a sizable portion of his boss's constituency but were nothing more than agitators and activists—most of them probably liberals.

That's when the first student spoke up. She shared how she had grown up in a conservative home and community, and that in many ways she maintained many of the values she had been taught. However, she said, she could not claim conservatism or the Republican Party for herself because of how thoroughly both had left her behind on climate change. Another student echoed her story, and soon similar stories were pouring forth from almost every other student in the room. I watched the chief of staff's face the entire time as his smile turned from a smirk to a grimace.

My face, on the other hand, beamed with pride. These students, most of them in their late teens or early twenties, were embodying the best of Spirit-shaped Christian citizenship. They stood inside the office of one of the most powerful people in the country and openly disagreed with him and his staff, yet did so without vitriol or hate but with kindness and humility. Their passion and resolve were not diminished by their gentleness but enhanced, the moral force of their message filling every corner of the room. The defensive smugness of the chief of staff evaporated as it became clear that he need not steel himself for a fight but need only open his ears and his heart to receive the gift these students offered him.

The students demonstrated that we can still advocate strongly for justice and righteousness in response to the climate crisis. But, like the priest, we do so from a place of spiritual rootedness and wholeheartedness. Our advocacy remembers that our true role in the work is always subordinate to Christ and as such must look like Christ. Our climate advocacy may be decidedly less headline-grabbing, but it becomes a whole lot more faithful.

I thought about this moment in the US Capitol four years later as I watched in horror as the same building was visited again by a group of citizens, many of them claiming Christ themselves. This time, though, the Christians looked different. Unlike the students who led with love, kindness, and goodness, these Christians were gripped by fear. This group of Christians had been willfully misled and had become confused about their role in addressing a perceived injustice. They believed the lie, so enticingly whispered in their ears by the powers and principalities of the day, that they had no other choice than to trade in the tools of Christ for the tools of violence. On January 6, 2021, in the same building where I had once watched Christian citizenship at its best, Christians showed the world what happens when they trade in the cross for the sword.

Christian Citizenship in a Warming World Is Grounded in Community

In the waning days of 2021, as global leaders huddled in Glasgow, Scotland, for the annual UN Climate Change Conference (COP26), some friends from my church and I huddled in the gathering night to light candles and pray for climate action. It was a small act of prayerful resistance against the deepening darkness of climate chaos, and we knew it. But we also knew that we needed it. At least, I needed it.

It came at a time in my life and my work when I was feeling like I was running on fumes. We were in the middle of an advocacy slog that felt never-ending, with a pair of bills working their way through Congress that, together, promised to be the single largest climate investment in US history. Yet, as we gathered in our pools of candlelight that night, their passage was as elusive as ever.

What was more, my wife was due a mere two months later with our second child. While I was overjoyed at the prospect of another little dear one to love, I was also feeling the existential weight of climate grief more than ever. I already mourned daily for the life my older son would live and the set of precarious circumstances that would compose his "normal." Now I had an entirely new innocent life to worry about—an entirely new set of circumstances and challenges for which to grieve.

I had gone to therapy for the first time earlier that year to work through it all, and part of that work helped me understand that I needed more opportunities to enact hope in the midst of the climate crisis with other people. I began to realize that the writing and campaigning and speaking and teaching that took up so much of my time was all necessary and important work, but that it all too often occurred in isolation. This was especially true after eighteen months of Covid-19 pandemic travel restrictions had migrated all of my work online.

I was lonely. I was running on empty. I couldn't hold hope for myself much longer. I needed a community to hold it for me. So, I gathered with other believers who shared my grief, who felt the threat of the climate crisis in their bones, just like I did. We gathered to lament, to pray, to grieve, and to hope.

I can't count how many young Christians I've met over the years who tell me how lonely they feel in their climate work. The conversations happen over coffee or in their campus cafeteria or at the front of a stage after I've given a talk. They tell me how heavy is their lament for the ways in which their gospel-rooted concern for climate change has separated them from their families. They share how isolated they feel in their passion for climate action on their campus, in their church, or even among their friends.

My advice to them is usually pretty simple: find your people. You can't do the work without them. It's just too lonely, too isolating, too hard on our own. Their response is usually simple too. How? How do we find our people in the work of climate action?

It's normal for those of us engaged in climate action to go through periods of loneliness and isolation. But at a certain point, isolation becomes self-perpetuating. If we convince ourselves that no one shares our heart, no one understands the depth of our climate grief, or that no one would want to walk the journey of climate action with us, then sooner or later that becomes our reality. We indulge in exercises of self-pity that, while cathartic, serve to isolate us even more. More often than not we already have people (or, at least, a person) who are in the fight with us. So, make a list of the people you already have in your life that share your commitment to faith-rooted climate action. Name them, and then send them a text to let them know how grateful you are for them.

Once you open your eyes to your own people already in your midst, find more people by showing up. There are countless opportunities to attend climate rallies, workshops, lectures, community organizing meetings, and marches. During the Covid-19 era, these events have only proliferated as the world has become fluent in Zoom and other online organizing formats. I found my people by getting out of my comfort zone. I jumped in a van with mostly strangers and drove to Washington, DC, joined the student environmental club on my campus, and volunteered with groups like Young Evangelicals for Climate Action. By pushing myself just a little, I found my people. As one of my people, Melody Zhang, puts it, I "put my body where my heart is," and to my surprise, I found lots of other beautiful hearts there too.

Your people may be from your church. They may be friends from college or high school. They may be close family members. They may be strangers-turned-friends that you met at that climate rally you pushed yourself to attend. Whoever they are, they are out there, and most of the time they are closer than you may think.

In the run-up to the 2020 election, I was working for Young Evangelicals for Climate Action. We decided to create a space where people who shared a Christian concern for climate action could take action together. We set up a bunch of Zoom calls and put the word out that we would be texting other young folks about the importance of voting in the November election with the climate in mind. We had a blast. People who had never done anything with us before signed up; then they signed up again and again. We shared funny stories. We laughed at some of the more memorable responses we would get back as we sent our texts. We sent almost two hundred thousand text messages, and we found joy in taking action together.

Be intentional about finding your people, and then be just as intentional about finding ways to take action together. Go to that climate rally together. Set up a Zoom room and write advocacy letters to your members of Congress together. When successes come, celebrate them together. When setbacks come, lament them together. Climate action is just too hard by yourself. Find your people, and do it together.

Christian Citizenship in a Warming World
Is Anchored in Spiritual Practices

Once you find your people, get to work grounding your individual and collective action in the deep, rich soil of spiritual practice. Anchoring our climate advocacy in spiritual practice isn't about checking a piety box or ensuring our faith doesn't slip into works righteousness by sprinkling in some prayer here and there. It's about uniting our heart with God's heart for his world, our vision with God's vision for creation's coming good future, and our action with God's saving and liberating action in the world. It's about making our climate advocacy more effective, more sustainable, and more Christlike.

Action rooted in spiritual practice is the reason Martin Luther King Jr. was able to sustain the nonviolent power of his movement even as the full force of White supremacy's demonic rage rained down on him and other civil rights activists. Action grounded in the reality of the kingdom of God is the reason Cesar Chavez was able to survive hunger strikes. It was the reason the migrant farm workers movement was able to overcome the violent tactics of powerful corporate interests and secure dignity and justice for workers through peaceful strikes and pilgrimages.

This is what makes Christian advocacy so powerful. Anchoring our climate action in spiritual practice decenters our efforts and centers the Spirit. It insulates us from the pendulum swing of elation and despair inherent to the work of engaging the climate crisis in the public square. It protects us from the temptation to resort to violence or coercion if our efforts are not succeeding, and from the risk of laziness and complacency if our efforts are yielding results at a given time. It strengthens us for the lifelong work of faithful advocacy.

More than this, grounding our climate action in spiritual practice situates our advocacy in a reality that transcends the one that we are trying to influence. It steeps us in the reality of the kingdom of God, and it entwines our will and our actions with the will and actions of the Holy Spirit who is, after all, the one doing all the work.

What these spiritual practices are and how they are practiced will look different for each Christian climate advocate. They may include deep prayer and Scripture study. They may be focused on meditation, mindfulness, and contemplation. They may look like many of the spiritual practices of climate action discussed in chapter seven and appendix A. You may practice them almost entirely on your own, within a worshiping community, or with some of the people you've found with whom you can walk the road of Christian climate action.

My discovery of a particular spiritual practice has been transformative for my own climate advocacy, and it supported me during a period when I really needed it. The "practice of the presence of God" is attributed to a seventeenth-century French monk named Brother Lawrence. Brother Lawrence was the cook at his monastery, and he had become renowned throughout the region at the time for his effervescent joy and contentment. People would pilgrimage from across the continent to meet him and would ask how it was that a man who filled his days with quotidian labor could be so sublimely wise and at peace.

The answer lay in a particular kind of mindfulness Brother Lawrence had developed that allowed him to focus on the presence of God all around him at any given time. Brother Lawrence believed that God's presence is radically imminent in creation, and that it is not God who hides himself from us but we who hide ourselves from God. By attuning our minds to God all around us, taught Brother Lawrence, we can experience communion with the divine even as we peel potatoes or scrub dirty pots and pans.

So, as he chopped carrots, he was marveling at the goodness of a God who would see fit to create such a gift as a carrot. As he stoked the fire, he would offer silent thanks for the brilliant design of fire, which offered warmth, comfort, and light. As he sloshed the dish water outside the back door of the kitchen, the wind would play against his cheek, and he would feel God in the touch of air against skin.

I learned about this particular practice only months before I traveled to Paris for the COP21 UN Climate Conference in 2015. The two weeks I spent there were mentally, emotionally, and physically exhausting. I walked miles every day (in dress shoes and a suit, no less), woke up early, and went to bed late. I even woke up one morning at two to host a live webinar update for people following our work back in the United States. When I look back on my journal from those two weeks, it is a roller coaster of highs and lows. But practicing the presence of God and engaging in other daily spiritual practices were constant steadying forces during that time. I practiced looking for God's presence on my train ride into the city every morning, while walking the halls of the convention center, and on my walk from the train station back home for the day. Each time I did, my breathing slowed and my buzzing brain cleared. These practices centered and sustained me. And my advocacy those two weeks was much more effective because of it.

I must admit that before this experience, I felt that this kind of contemplation and meditation was pretty out there. At the very least, it was for those privileged enough to be able to escape the world while nitty-gritty action was for those committed to changing the world for the better. Contemplation was at best indulgent navel gazing and at worst a willful rejection of those suffering oppression all around us who were crying out for justice.

Yet, over time I've come to understand the intimate relationship between spiritual practice and action. I've come to recognize that a focus on either one at the expense of the other is dangerous. I've realized that focusing solely on spiritual practices in the name of moralism or piety at the exclusion of action is, indeed, indulgent, privileged, and unfaithful. Likewise, focusing exclusively on action and ignoring the spiritual practice that fuels and motivates action is self-aggrandizing and dangerously unsustainable. The equation for faithful Christian citizenship in a warming world, I have come to learn, is spiritual practice + action. Neither is complete without the other; neither is effective without the other.

Christian Citizenship in a Warming World as Vocation

In many ways, we can think of faithful Christian citizenship in a warming world as one part of our larger Christian vocation. Vocation is often equated simply with career. Sometimes we may even be under the impression that only those called to the "sacred" professions of ministry have a vocation while the rest merely muddle through their "mundane" nine-to-fives. This couldn't be more wrong.

The truth is, each of us has a vocation. And rather than simply comprising a career, each of our vocations comprises a vast constellation of callings that are shaped by the joys, fears, dispositions, sorrows, and desires that are beautifully particular to each of us. No constellation of callings is exactly the same. The shape of your vocation is as unique as your fingerprint.

Frederick Buechner, a Christian pastor and author, put it more succinctly: "The place God calls you to is the place where your deep gladness and the world's deep hunger meet."[6] I love Buechner's insight on both the joy and the need of vocation. Our lives are meant to embody good news to the poor, especially in a world being rocked by the life-and-death consequences of a changing climate. And we are meant to find fulfillment in the process. Yet, as apt as his observation is, I humbly submit (with more than a little fear and trembling) a friendly amendment to his definition: "*The places* God calls you to *are the places* where your deep gladness and the world's deep hunger meet."

If vocation is a constellation of callings, then we are called to myriad places and practices that marry our deep gladness and the deep hunger of the world. The ways in which we consume goods and services in a globalized world, the relationships we commit ourselves to, and—yes—how we make a living are all callings that, together, comprise our vocational constellation.

However, as unique as each of our vocations is, I hope I've made the case throughout this book that every one of us has a star cluster in our constellation called "climate action." The witness of Scripture, a full-throated gospel, and a true defense of the sacred gift of life all mean that care for God's creation and all who depend on it is an irreducible part of every Christian's vocation. In a warming world, this means every single Christian has a vocational responsibility to respond to the climate crisis with faith, hope, and love. It's just part of what it means to follow Jesus in the twenty-first century.

And because each of our vocations is utterly unique, how we live this out will look different for each one of us. The story that you have to tell of your personal relationship to creation and your particular reasons for addressing climate change will be utterly your own. The array of disciplines that you incorporate into your life in order to be better formed into the image-bearing protector of creation you are called to be will look like nobody else's. The various actions that you undertake to live out your climate calling in the public square will constitute an advocacy diet all your own—each a beautiful, irreplaceable star in the larger constellation of your life.

In other words, caring for creation and addressing the climate crisis is common to the vocation of every follower of Jesus. Yet, how each of us lives out this common vocational calling in the world is as singular as each one of our wild, precious lives. So, how are we to discover the distinct ways in which each of us is meant to live into this common call? A Quaker saying offers simple advice: "Let your life speak." Author and activist Parker Palmer reflects on this maxim and observes, "Vocation does not come from willfulness. It comes from listening."[7] That constellation of callings that together we discern as our vocation (and that particular star cluster called "climate action"), says Palmer, is not "a goal to be achieved but a gift to be received."[8]

We often think of vocation as some secret master plan for our lives that God drew up before the beginning of time and is now hiding from us, challenging us to discover it. The anxiety I've witnessed from young people striving to decipher their one true vocation before time runs out is as pervasive as it is heartbreaking. Vocation is meant to be a gift. It is the sum of all the beautiful, quirky, unique ways in which God has designed you to find joy as you join him in the work of proclaiming the gospel of his coming kingdom to a warming world desperate for good news.

Author and English professor Debra Rienstra captures this well in her book *So Much More: An Invitation to Christian Spirituality*. She compares the task shared by a mother and her young son baking cookies to the partnership between us and God in the task of proclaiming redemption to God's broken world. In Rienstra's telling, the kid spills flour, measures ingredients inexactly, and sneaks handfuls of chocolate chips when no one is looking. Mom has to follow behind him, cleaning up his messes. As Rienstra observes,

> It's a terribly inefficient operation. But it has value other than efficiency, in teaching the child and in the loving companionship built by a shared task. I imagine God would sometimes like to shoo us out of the way and get down to business without our help. But like a wise mother, God generously welcomes us again and again back into the kitchen.[9]

Bishop Ken Untener understood the same basic truth about our vocational task and captured it in his beautiful prayer *Prophets of a Future Not Our Own*:

> The kingdom is not only beyond our efforts, it is even beyond
>   our vision.
> We accomplish in our lifetime only a tiny fraction of the
>   magnificent enterprise that is God's work. Nothing we do
>   is complete, which is a way of saying that the Kingdom
>   always lies beyond us.
>
> . . . .
>
> This is what we are about.
> We plant the seeds that one day will grow.
> We water seeds already planted, knowing that they hold
>   future promise.
> We lay foundations that will need further development.

We provide yeast that produces far beyond our capabilities.
We cannot do everything, and there is a sense of liberation in
realizing that.
This enables us to do something, and to do it very well.
It may be incomplete, but it is a beginning, a step along the
way, an opportunity for the Lord's grace to enter and do
the rest.[10]

This is the task of each of us striving to live out our vocations as we follow Jesus in a warming world: to see the suffering of God's groaning creation, to resist the temptation to look away, and to let it move us to play our own small part in the grand story of God's redeeming work in the world. And just when we begin to fool ourselves into thinking that the task of solving climate change is completely up to us, we will drop another eggshell in the batter. And God, following close behind, will patiently pick it out.

COMMENCEMENT DAY, 2066

May 22, 2066

Dear One,

It's funny, isn't it, how we humans make meaning? How dots and lines on a page—even this page—can unlock pain and joy in equal measure. How something as small as the wink of an eye can wrap us in warmth and belonging. How a bite of bread and a swallow of juice consumed with others can somehow catch us up, if even for an instant, in the mystery of heaven. How sprinkling water over your downy head eighteen years ago filled this old man's heart to bursting.

And how, in a few short hours, a brief walk across a stage will usher you into an entirely new world of possibility and potential. On this your day of commencement, dear Granddaughter, I hope you'll suffer an old man's brief sentimentality before bounding into your glorious future.

On the day you were born, I spent a lot of time ruminating on your future. An old habit. I guess I picked it up around the time your dad was born. Back then, the fierce storms and punishing heat that have been taken for granted during your life were just beginning to break through into our reality. For much of my life to that point, they had been mostly abstractions—dangerous offspring of our inaction that would one day grow up and move out of the house to wreak their havoc on the earth but innocuous enough as they merely gestated in the womb of our collective ignorance and denial.

By the time your dad was born in 2018, though, the consequences of our procrastination were becoming harder and harder to ignore. There were some our age, even then, who were choosing not to have kids. Deciding that the future was too dangerous, too unpredictable to be able to morally justify yoking a human life to it for decades to come without that human's prior and informed consent, a sentiment your grandma and I could certainly understand, though never quite embrace. I guess our hope in God's good plans for the world has always been more stubborn than our fear of our ability to derail them. But that doesn't mean the fear hasn't been there, ever mingling with the hope.

On the day your father came into the world, that alloy of hope and fear was forged and lodged deep in my heart for good. There's a paradox to loving other mortals, that even as your heart remains fixed in your chest, its twin beats inside someone else's. You watch your own heart's mirror as it jumps and laughs and aches. It's a phenomenon that repeats itself whenever we make the dangerous, awesome choice to love. All these years, as my own fearful heart has pumped dutifully inside my aging chest, it has replicated itself as first your dad and his siblings were born and then again when you and your siblings and cousins all entered this precious, precarious place. All of my Dear Ones.

I still remember each of those days perfectly, you know, including yours. That first time I held you in my arms—my heart's newest match. I admit that in that moment my heart was mostly fear. Don't get me wrong: we had made significant progress at pulling the world back from the brink. Countries had finally dug in—decades later than they should have—and made many of the necessary, difficult decisions that needed to be made.

The acres of solar panels and seas of windmills that you take for granted were erected in earnest. The jobs that had evaporated from traditional manufacturing hubs like Detroit, Akron, and Pittsburgh came flooding back as the demand for electric vehicles—and all their component parts—skyrocketed. Investments in high-speed public transit connected the country in ways not seen since Eisenhower latticed the nation with interstates. I know it may be hard to believe, but you weren't always able to take that high-speed train from Iowa City to come see your grandma and me in Michigan and make it home again by dinner.

Environmental atrocities that were allowed to persist for generations were finally addressed head-on. Lead pipes were exhumed, wide-open oil and gas wellheads spewing methane to the heavens were finally capped, and toxic coal ash was contained. Black and Brown lungs and brains were allowed to heal at last.

Monopolistic electric utilities gave way to community energy as neighbors banded together to put up solar panels on their roofs and to share the electricity with one another. Noise pollution everywhere plummeted as the internal combustion engine swapped out its place under the hood for a museum plinth, bringing migrating birds back over major cities and marine life back to their traditional migration routes. Empowered by new agricultural policy, an entire generation became reacquainted with the intimate interplay of soil and seed. Food began to be grown drastically closer to home for the first time in a century. And, thanks be to God, the church finally began to claim its rightful place in the public square, demanding these measures in the name of neighbor-love and the common good. Demanding them in God's name.

You've learned enough in your history classes to know that your experience of church is drastically different from what mine has been for most of my life. You know that the way most churches today transition seamlessly on a Sunday morning from worship in the sanctuary to marching in the streets used to be alien. That all the polls that consistently show Christians at the top of the list of people in society most concerned about the environment and climate change is a transformation that even I dared not dream of decades ago. That churches blanketed with solar panels, dotted with community gardens, and swaddled in swaths of native plants and grasses—so commonplace today—were exotic anomalies for much of my lifetime.

Indeed, it would be hard for you to understand how fully the church and the Christians who constitute it have been transformed in the handful of decades that came before today. This day. Your special day.

I'm sorry to say that, would you somehow be able to travel those decades in reverse and find yourself in a church all those years ago, you might not recognize it. Simply put, the church has come back to itself. There is hope, my Dear One. So much hope.

So why does the fear remain? Perhaps it's an old man's feeble heart that still breaks for a future it couldn't secure for you. We knew all those years ago what our slow walking and half measures would mean for you. We knew that even with all the progress we have made over the last several decades, powerful climate impacts were already baked into the atmosphere. We knew that your world would be fundamentally altered from the one we had known. We knew that glorious future you are getting ready to explore would be more dangerous and more unpredictable because of us.

Maybe it's a fear that can't let go of the world it used to know enough to find all the beauty and joy that are still to be had. Maybe it's a fear that's grown too familiar over the years and can't be shed—at least not entirely. Maybe it's a fear that, though you never have, you still might lock me with your stunning stare and I'll find accusation in your eyes. Maybe it's a fear that, though I tried to do all I could with what I knew, it still wasn't enough—that I couldn't be enough for you.

Whatever the reasons, the fear remains, mingled with the hope. And that's okay. I've come to learn across these many years that hope and fear, like twinned hearts, are bound together. And though there were times when I thought the fear in my heart would swallow up the hope, it has not. In fact, the hope has been the half of my heart that has stretched and grown the most as the years have passed. Even though fear swelled in me as I held your newborn body against mine all those years ago, it's hope that I feel more than anything else today. And I have you, Dear One, to thank for that.

It's funny how we humans make meaning. A tassel flipped from one side to the other. A hat thrown in the air. This beautiful, wild, precious world is bursting at the seams with meaning. So go, Dear One, and make it.

Make meaning in the dappled sunlight streaming through waving leaves. In the beauty of birdsong and rushing wind. In charging waves and the utter stillness of night. In the glory and presence of the Creator pulsing through it all. In choosing to yoke your heart to another, in all its exhilarating gift and terrifying risk. In the mysterious alchemy of hope and fear.

And though it may feel at times that your heart just might be swallowed up by that fear, know this: there was once a man who thought the same thing. He thought that the fear for your life and your future would blot out all else. Instead, he has spent every day of the past eighteen years astounded by his feeble heart's ability to love you more than it did the day before. And that love has made all the difference.

All my love, Dear One, on this day and every glorious day to come.

Grandpa

ACKNOWLEDGMENTS

THIS BOOK IS ONLY IN YOUR HANDS (or on your ebook device or being read through your headphones) because countless people believed in it and in me, and pushed me to do the hard work to get it out into the world.

It would be impossible to thank everyone that deserves to be. If I forget you, forgive me. And know that whether or not your name appears on these pages, you made this book possible. I am forever grateful to you.

This book never would have happened without the dedicated, capable, and brilliant team at InterVarsity Press. Justin Paul Lawrence recognized me right away at Urbana 2018 after we hadn't seen each other in years and told me that my dream to write this book wasn't crazy. Thanks to everybody on the marketing and sales teams who worked like crazy to give this book its best shot in life.

A huge thank you to my editor, Ethan McCarthy, who gets his own paragraph for putting up with my obsessiveness and my endless email threads, and who alone knows how much work my early drafts needed. Ethan, your gentleness with my words, your firm and wise counsel, and your dogged belief in this book have meant more to me than you know.

Special thanks to all of my friends, colleagues, and coconspirators at Young Evangelicals for Climate Action and the Evangelical Environmental Network: Mitch Hescox, Jessica Moerman, Marqus Cole, Kim Anderson, Tori Goebel, Lindsay Garcia, and Jeremy Summers. To all of the student leaders who have passed through YECA over the years and whose stories and experiences pepper these pages: you all inspire me daily to work smarter, to love greater, and to hope despite the headlines.

I owe a major debt of gratitude to Larry, Margaret, and Robert, who so freely and generously offered me the sacred gift of their stories. I hope I've done them justice and that I've proven myself worthy of the gift.

There were also friends who stepped in at crucial moments to offer hospitality, expertise, and wisdom that have made this book infinitely better: Kristin Kobes Du Mez, who offered invaluable insight into early pages and made sure I didn't embarrass myself too much with my historical work in chapter two; Travis West, who answered the eleventh-hour call to offer his brilliant expertise into the profound depths of the Hebrew language; and Jon Terry and Eric Bond at Au Sable Institute, who offered me space rent-free for a writing weekend at a critical stage in the early writing process. That Northern Michigan lake cottage in late August was both respite and muse.

My brother, Brian Schaap, started it all for me and has never once told me to stop dragging him into it. For all those late-night talks and the constant companionship, I am so grateful.

I am always thankful for Brian Webb, John Elwood, Shay O'Reilly, Chris Elisara, Ben Lowe, Kermit Hovey, Ed Brown, Lowell Bliss, Rich Killmer, Peter VanderMeulen, Peter Fargo, Steve Mulder, Kris Van-Engen, Andrew Oppong, Debra Reinstra, Peter Illyn, Anna Jane Joyner, Dave Warners, Gail Heffner, Matt Heun, Rusty Pritchard, Jason Fileta, Sarah Withrow King, and all my other friends in the evangelical and evangelical-adjacent climate movement. Your partnership in the work means the world to me.

Finally, I owe the most to my wife and partner, Allie. Allie always believed that I could write this book, even and especially when I had my doubts. She always believed that the world needed to hear what I had to say, even and especially when my inner critic said otherwise. And she acted like it. She spent evenings putting kids to bed by herself while I wrote. She did my share of the dishes, extra laundry loads, and meal planning while I wrote. She did weekends as a solo parent while I wrote. Allie, I will never be able to repay you for the gift you've given me. This book, and my heart, belongs to you.

This book is dedicated to my kids, current and future, and to their potential future kids. I hope I did them proud.

Appendix A

THE SPIRITUAL DISCIPLINES OF CLIMATE ACTION

THERE ARE SO MANY WAYS that we can align our values with the ways we live our lives to better care for God's creation and to better love our neighbors. The following list is by no means meant to be exhaustive, but rather illustrative, to spark your imagination and to give you a toehold to begin doing something, right now.

Some might read this list and get started on all of them at once. I wouldn't recommend it. Instead, prayerfully consider one or two new practices and try them out. Since each of our vocational constellations is unique, each of our distinct callings to practice the spiritual disciplines of climate action will be too. I don't know the particular ways that God is calling you into deeper faithfulness through these practices—but he does. Approach them in prayer, try a few on, discard the ones that don't fit quite right, and try some more. And remember: expect joy!

The Spiritual Practice of Climate Action: An Incomplete List

- Calculate your carbon footprint at Global Footprint Network (www.footprintcalculator.org/home/en).
- Change lightbulbs to LEDs, which use 80 percent less energy than incandescent and last twenty-five times longer.
- Use ENERGY STAR lights, appliances, etc.

- Turn your thermostat up in summer and down in winter, especially if you are not home. Even a couple of degrees can make a big impact!
- Turn off lights and unplug devices when not in use.
- Run only full loads in the dishwasher or washing machine.
- Use only cold water to wash clothes.
- Line-dry clothes.
- Purchase renewable energy from your utility provider.
- Take advantage of energy-efficiency programs through your utility provider.
- Install solar panels.
- Better seal and insulate your home.
- Drive less often.
- Keep tires properly inflated.
- Carpool, walk, or bike when possible.
- Use public transportation.
- Refuse, rethink, repurpose, reduce, reuse, recycle (in that order!).
- Purchase durable products that will last over disposable ones.
- Research the ecological and social impacts of your purchases (try betterworldshopper.com).
- Plant a vegetable garden or purchase a plot at a community garden.
- Bring reusable bags to the grocery store.
- Wash and reuse single-use plastic bags.
- Use reusable water bottles.
- Compost food waste.
- Eat less meat, especially beef.
- Buy local products.
- Purchase clothes from secondhand stores rather than new.
- Challenge yourself to buy all your produce next summer from the farmers' market.

- Buy a produce share from a local CSA (Community Sponsored Agriculture).
- Fly less often—especially internationally.
- When you do need to travel, purchase carbon offsets for travel and lifestyle emissions (try https://climatestewardsusa.org/offset/).
- Tell your members of Congress to support serious climate policies.
- Vote for candidates who are committed to enacting strong climate policies.
- Attend a climate protest/demonstration.
- Post something on social media about why you are concerned about climate change and what you are doing to respond to it.
- Tell a friend or family member your climate story.
- Keep learning.
- Stay informed.
- Share these personal commitments with others.

Appendix B

WRITING FOR CHANGE

OP-EDS ARE A POWERFUL WAY to advocate for climate action. Below are more details on that framework for a successful op-ed we discussed in chapter eight, along with examples from op-eds I've been able to get published throughout the years:

- **Lede:** *Lede* is news jargon for the first one to three sentences of a story—in this case, an op-ed. It is the opening that grabs readers by the shirt collar and convinces them that its worth their time to keep reading. It is typically organized around a timely news hook, whether recent breaking news, the approach of a new season or holiday, or the release of a major new study.

  - For example, "Much of the American West is on fire, Atlantic hurricanes have exhausted the alphabet two months early, and along the way climate change has once again muscled its way back into the national conversation."

- **Thesis:** The thesis is the argument an op-ed is trying to advance. Unlike with books, dissertations, or term papers, the thesis in an op-ed can be either explicit or implicit.

  - For example, "For this, we largely have young people to thank. Among this growing throng of youth climate activists are some you might not expect: young evangelical Christians."

- **To Be Sure:** A "to be sure" is a qualifier or caveat to the argument being advanced in the op-ed. This is a staple of the genre, and at least one must be included in any op-ed hoping to be published. These anticipate counterarguments to the thesis you are advancing and offer a thoughtful rebuttal. This signals to both the reader and the editor deciding whether to publish a piece that the author is thoughtful, humble, and not dogmatic—this person can recognize that their argument isn't perfect.

- For example, "Evangelicals in the United States are hardly leading the charge for climate action. Polls consistently show that White evangelicals remain the group that is most skeptical of the science behind human-driven climate change and least receptive to proposed solutions. . . . Yet these numbers are a lagging indicator, and what they fail to capture is that young evangelicals are becoming overwhelmingly supportive of the need to address climate change."

- **Evidence-Based Arguments:** A well-written op-ed should include two to three pieces of evidence that support the thesis. This evidence can be in the form of stats, polls, surveys, news reports, studies from credible organizations, expert quotes, scholarship, history, or even firsthand experience. Hyperlink to primary sources as much as possible.

  - For example, "Though our parents and grandparents have dictated evangelical political prerogatives in the past, millennials and Gen Z are ascendant. Almost 40 percent of eligible voters in 2020 will belong to these forty-and-under generations, according to Pew" (study from a credible source).

  - For example, "This kind of data supports what I've been seeing on the ground over the last several years. Namely, that there is a tectonic shift taking place in the church right now, and young Christians are at its epicenter" (firsthand experience).

- **Conclusion:** Like all good argument-driven writing, an op-ed should wrap up with a clear, succinct conclusion. Sometimes the conclusion will include a call to action for readers who want to respond to the argument being advanced. Sometimes it will circle back to the lede to form a bookend to the piece.

- For example, "For Christians, safeguarding God's works and protecting vulnerable people from climate pollution are both invitations to get better at following Jesus—the Jesus who said that nothing was more important than loving God and loving our neighbors."

Pitching your op-eds is as important as writing them. Here are those three questions every editor will ask when they receive your pitch, with a little more detail:

- **Why me?** Editors will ask themselves why you are uniquely qualified to be making the argument you are making. Make sure to lean into any bona fides you might carry, whether titles, awards, or other published pieces. Show editors that you are a leader whose voice carries weight because, if you are a Christian engaged in climate action work, you are and it does!

- **Why now?** Editors are always looking to peg reader-submitted pieces to the news cycle. They will approach your piece with the question, "Why do our readers need to hear this argument right now?" Make the case for why your piece is timely and important.

- **So what?** Editors will need to be convinced that your argument is necessary for their readers to hear. Help them understand the stakes of the moment and how your argument advances the public discussion in an essential way that no other argument can or has.

NOTES

Introduction

[1]Unless noted, when speaking about evangelicals I am referring to the White, politically conservative subgroup that dominates American evangelicalism today. While a 2014 Pew survey found that 24 percent of American evangelical Protestants are non-White, most social, religious, and political expressions of American evangelicalism have been shaped by White, conservative identity.

[2]John Cook et al., "Consensus on Consensus: A Synthesis of Consensus Estimates on Human-Caused Global Warming," *Environmental Research Letters* 11, no. 4 (April 2016): 048002, http://dx.doi.org/10.1088/1748-9326/11/4/048002.

[3]Kimberly Nicholas, *Under the Sky We Make: How to Be Human in a Warming World* (New York: Putnam's Sons, 2021).

[4]Leiserowitz et al., "Climate Change in the American Mind: April 2020," Yale Program on Climate Change Communication, https://climatecommunication .yale.edu/publications/climate-change-in-the-american-mind-april-2020/.

[5]Stephan Lewandowsky, Giles E. Gignac, and Samuel Vaughan, "The Pivotal Role of Perceived Scientific Consensus in Acceptance of Science," *Nature Climate Change* 3 (October 2012): 399-404, https://doi.org/10.1038/nclimate1720.

[6]Paris Agreement, United Nations, 2015, article 2.1(a), 3, https://unfccc.int/files /essential_background/convention/application/pdf/english_paris_agreement.pdf.

[7]Paris Agreement, article 2.1(a), 3.

[8]V. Masson-Delmotte et al., "IPCC, 2018: Summary for Policymakers," in *Special Report: Global Warming of 1.5°C* (New York: Cambridge University Press, 2018), section C.2, 15, www.ipcc.ch/sr15/chapter/spm/.

[9]Jamal Srouji et al., "Closing the Gap: The Impact of G20 Climate Commitments on Limiting Global Temperature Rise to 1.5°C," World Resources Institute and Climate Analytics, September 2021, 4, www.wri.org/research/closing-the-gap -g20-climate-commitments-limiting-global-temperature-rise.

[10]"Temperatures," Climate Action Tracker, last modified November 9, 2021, https:// climateactiontracker.org/global/temperatures/.

1. Coal and the Greatest Commandment

[1]"What Is Mountaintop Removal Coal Mining?," iLoveMountains.org, accessed June 22, 2022, https://ilovemountains.org/resources.

2. How Did We Get Here?

[1]Gillian Frank and Neil J. Young, "What Everyone Gets Wrong About Evangelicals and Abortion," *Washington Post*, May 16, 2022, www.washingtonpost.com /outlook/2022/05/16/what-everyone-gets-wrong-about-evangelicals-abortion/.

[2]Randall Ballmer, *Bad Faith: Race and the Rise of the Religious Right* (Grand Rapids, MI: Eerdmans, 2021); Katherine Stewart, *The Power Worshippers: Inside the Dangerous Rise of Religious Nationalism* (New York: Bloomsbury, 2019).

[3]Darren Dochuk, *Anointed with Oil: How Christianity and Crude Made Modern America* (New York: Basic Books, 2019), 12.

[4]Dochuk, *Anointed with Oil*, 46.

[5]Dochuk, *Anointed with Oil*, 509.

[6]Ronald Reagan, "National Affairs Campaign Address on Religious Liberty (Abridged)," delivered August 22, 1980, American Rhetoric, www.americanrhetoric .com/speeches/ronaldreaganreligiousliberty.htm.

[7]In the age of climate chaos and mass extinction, I can't help but wonder which sin grieves God's heart more—too much love for creation or too little? If only our great sin were that we loved creation too much.

[8]More on that word *new*—and how it doesn't mean "from scratch"—in chap. 3.

[9]N. T. Wright, *Surprised by Hope: Rethinking Heaven, the Resurrection, and the Mission of the Church* (San Francisco: HarperOne, 2008), 148.

[10]John MacArthur, "The End of the Universe, Part 2," *Grace to You*, September 21, 2008, www.gty.org/library/sermons-library/90-361/the-end-of-the-universe-part-2.

[11]Stanton A. Glantz et al., *The Cigarette Papers* (Berkeley: University of California Press, 1996), 190.

[12]"Smoke & Fumes," Center for International Environmental Law, 2016, www .smokeandfumes.org/fumes.

[13]Naomi Oreskes, Erik M. Conway, *Merchants of Doubt* (New York: Bloomsbury, 2010).

[14]"Religious Identities and the Race Against the Virus: American Attitudes on Vaccination Mandates and Religious Exemptions (Wave 3)," PRRI, December 9, 2021, www.prri.org/research/religious-identities-and-the-race-against-the-virus -american-attitudes-on-vaccination-mandates-and-religious-exemptions/.

[15]"Competing Visions of America: An Evolving Identity or a Culture Under Attack," PRRI, November 1, 2021, www.prri.org/research/competing-visions -of-america-an-evolving-identity-or-a-culture-under-attack/.

[16]"Competing Visions of America" and "Understanding QAnon's Connection to American Politics, Religion, and Media Consumption," PRRI, May 27, 2021, www.prri.org/research/qanon-conspiracy-american-politics-report/.

[17]"Religion and Views on Climate and Energy Issues," Religion and Science, Pew Research Center, October 22, 2015, www.pewresearch.org/science/2015/10/22/religion-and-views-on-climate-and-energy-issues/.

[18]"National Poll TopLines," Interfaith Power and Light National Climate Change Poll, October 13, 2020, https://climatenexus.org/wp-content/uploads/2015/09/IPL-National-Climate-Change-Poll.pdf. It's important to note that the Climate Nexus poll asked its question in a slightly different way from the 2015 Pew poll. Climate Nexus asked if respondents believed that climate change was caused mostly by human activity, as opposed to Pew's phrasing asking respondents if they believed climate change was caused by human activity. It is possible that the word *mostly* may be doing quite a bit of heavy lifting and can explain some of the discrepancy in the response rates.

[19]Matthew J. Hornsey et al., "Meta-analyses of the Determinants and Outcomes of Belief in Climate Change," *Nature Climate Change* 6 (February 22, 2016): 622-26, https://doi.org/10.1038/nclimate2943.

[20]Of course, where partisan political identity is entwined with religious identity, heresy and idolatry also flourish. The terrifying rise in the prevalence of Christian nationalism across the United States over the last several years, with millions of conservative Christians in its thrall, has shocked many. The rise of the heresy of Christian nationalism is less a shocking anomaly and more the logical conclusion of a movement dedicated to convincing millions of Christians that their religious identity and their partisan, political identity are the same thing. Why wouldn't those formed by such a movement come to blur the lines between Christ and nation? Between party and God?

[21]Marguerite Michaels, "Billy Graham: America Is Not God's Only Kingdom," *Parade* (February 1, 1981).

[22]"How the Faithful Voted: 2012; Preliminary Analysis," Religion & Politics, Pew Research Center, November 7, 2012, www.pewforum.org/2012/11/07/how-the-faithful-voted-2012-preliminary-exit-poll-analysis/; and "Religion and the Presidential Vote," Pew Research Center, December 6, 2004, www.pewresearch.org/politics/2004/12/06/religion-and-the-presidential-vote/.

3. Recovering the Big Story

[1]Portions of this chapter were previously published in Kyle Meyaard-Schaap, "All Things," Bible study, Young Evangelicals for Climate Action, accessed

October 24, 2022, https://d3n8a8pro7vhmx.cloudfront.net/een/pages/2055
/attachments/original/1557515470/All_Things_Bible_Study_2019_Updated
.pdf?1557515470; used with permission.

[2]Randy Woodley, "Randy Woodley on Indigenous Theology and the Harmony
Way," *The Luxcast*, episode 5.4, Western Theological Seminary, October 19, 2018.

[3]Parts of this section first appeared in Kyle Meyaard-Schaap, "The Word Became
Flesh," resources, *Reformed Worship*, September 2015, www.reformedworship
.org/article/september-2015/word-became-flesh; used with permission.

[4]Steven Bouma-Prediger, *For the Beauty of the Earth: A Christian Vision for Creation Care* (Grand Rapids, MI: Baker Academic, 2001), 105-10.

[5]The grammar of Colossians 1:15-20 bolsters the case for a face-value interpretation of "all things." Like many languages, Koine Greek is gendered. Unlike
English, where pronouns are gendered but common nouns are not, Koine Greek
applies a gender to people, places, and things. Yet, unlike some other gendered
languages, it contains three genders: feminine, masculine, and neuter. Every instance of the phrase *ta panta* in Colossians 1:15-20—along with the two other
forms that the adjective "all" takes in these verses—is neuter.

This is significant. Only twice in Colossians 1:15-20 does the adjective "all"
describe a specific noun: in Colossians 1:15, when Paul argues that Christ is the
"firstborn over all creation," and in Colossians 1:19, when Paul says that God was
"pleased to have all his fullness dwell in [Christ]." In these two instances, "all" is
feminine and neuter, respectively, to match the feminine noun "creation" (*ktisis*)
and the neuter noun "fullness" (*plērōma*). In these instances, the adjective "all"
is functioning descriptively. It is describing a noun, and therefore must match
the gender of the noun that it is modifying. The other six times that "all" is used
in this section, it stands alone. This means that it is functioning differently.
Rather than functioning descriptively, it is functioning substantivally—standing
in as a substitute for a noun. This means it is effectively functioning as a noun,
and subject words will often need to be added to it for clarity. In the case of
Colossians 1:15-20, that subject word is "things."

But why not "people" instead of "things"? In a word: gender. While there are
instances where Paul seems to use the neuter form of *pas* to refer to people
(Romans 1:16; 1 Corinthians 1:2; 2 Corinthians 3:2), there is much more evidence
throughout Paul's New Testament writings of him using the masculine form
when referring to his audience. Similarly, there is an abundance of examples of
Paul using the neuter form to refer to "everything" or "all things." Both the
grammar and Paul's own writing style confirm that Paul was after something
specific when he chose to repeat *ta panta* over and over again in

Colossians 1:15-20 when he refers to the Christ who created all things, holds all things together, and, crucially, reconciles all things to himself by making peace through the blood of his cross. He really seems to have meant *all things*.

[6]Palm branches likely obtained nationalistic significance for first-century Jewish populations because it was the symbol engraved on the last coins minted while the Jewish people living in Palestine were still free.

[7]John MacArthur, "The End of the Universe, Part 2," *Grace to You*, September 21, 2008, www.gty.org/library/sermons-library/90-361/the-end-of-the-universe-part-2.

5. Being Pro-Life in the Age of Climate Chaos

[1]"Climate Change: Regional Impacts," University Center for Atmospheric Research, accessed June 23, 2022, https://scied.ucar.edu/learning-zone/climate-change-impacts/regional.

[2]"Land Surface Temperature in the Sakha Republic," *Copernicus: Europe's Eyes on Earth*, June 21, 2021, www.copernicus.eu/en/media/image-day-gallery/land-surface-temperature-sakha-republic.

[3]Michaeleen Doucleff, "Anthrax Outbreak in Russia Thought to Be Result of Thawing Permafrost," NPR, August 3, 2016, www.npr.org/sections/goatsandsoda/2016/08/03/488400947/anthrax-outbreak-in-russia-thought-to-be-result-of-thawing-permafrost.

[4]Sam Tanzer, "Why Is There Such a Strong Correlation Between Geographic Distance from the Equator and Prosperity?," *Forbes*, March 20, 2012, www.forbes.com/sites/quora/2012/03/20/why-is-there-such-a-strong-correlation-between-geographic-distance-from-the-equator-and-prosperity.

[5]Simon Evans, "Analysis: Which Countries Are Historically Responsible for Climate Change?," *Carbon Brief*, October 5, 2021, www.carbonbrief.org/analysis-which-countries-are-historically-responsible-for-climate-change.

[6]C. Todd Lopez, "DOD, Navy Confront Climate Change Challenges in Southern Virginia," *DOD News*, US Department of Defense, July 21, 2021, www.defense.gov/News/News-Stories/Article/Article/2703096/dod-navy-confront-climate-change-challenges-in-southern-virginia/.

[7]Michael Batty et al., "Introducing the Distributional Financial Accounts of the United States," *Finance and Economics Discussion Series*, Divisions of Research & Statistics and Monetary Affairs, Federal Reserve Board, Washington, DC, 26, last updated January 9, 2020, https://doi.org/10.17016/FEDS.2019.017.

[8]Corinne N. Thompson et al., "COVID-19 Outbreak—New York City, February 29–June 1, 2020," *Morbidity and Mortality Weekly Report* 69, no. 46 (2020): 1725-29, http://dx.doi.org/10.15585/mmwr.mm6946a2.

[9]Jennifer Holleis, "How Climate Change Paved the Way to War in Syria," DW, February 26, 2021, https://www.dw.com/en/how-climate-change-paved-the-way-to-war-in-syria/a-56711650.

[10]"Health and Environmental Effects of Particulate Matter (PM)," EPA: United States Environmental Protection Agency, accessed June 23, 2022, www.epa.gov/pm-pollution/health-and-environmental-effects-particulate-matter-pm.

[11]Ian Parry, Simon Black, and Nate Vernon, "Still Not Getting Energy Prices Right: A Global and Country Update of Fossil Fuel Subsidies," *IMF Working Papers*, International Monetary Fund, September 24, 2021, www.imf.org/en/Publications/WP/Issues/2021/09/23/Still-Not-Getting-Energy-Prices-Right-A-Global-and-Country-Update-of-Fossil-Fuel-Subsidies-466004.

[12]Renee Skelton and Vernice Miller, "The Environmental Justice Movement," NRDC, March 17, 2016, www.nrdc.org/stories/environmental-justice-movement.

[13]Richard Rothstein, *The Color of Law: A Forgotten History of How Our Government Segregated America* (New York: Liveright, 2018).

[14]Rothstein, *Color of Law*, 127-28.

[15]"Ozone," American Lung Association, last updated April 20, 2020, www.lung.org/clean-air/outdoors/what-makes-air-unhealthy/ozone.

[16]"Oil and Gas Threat Map," Earthworks and FracTracker Alliance, accessed June 23, 2022, https://oilandgasthreatmap.com/threat-map/.

[17]Shaina L. Stacy et al., "Perinatal Outcomes and Unconventional Natural Gas Operations in Southwest Pennsylvania," *PLoS One* 10, no. 6 (2015): e0126425, doi: 10.1371/journal.pone.0126425; Lisa M. McKenzie et al., "Birth Outcomes and Maternal Residential Proximity to Natural Gas Development in Rural Colorado," *Environmental Health Perspectives* 122, no. 4 (2014), http://ehp.niehs.nih.gov/1306722; and Janet Currie, Michael Greenstone, and Katherine Meckel, "Hydraulic Fracturing and Infant Health: New Evidence from Pennsylvania," *Science Advances* 3, no. 12 (2017): e1603021, www.doi.org/10.1126/sciadv.1603021.

[18]Lilian Calderón-Garcidueñas et al., "Quadruple Abnormal Protein Aggregates in Brainstem Pathology and Exogenous Metal-Rich Magnetic Nanoparticles (and Engineered Ti-rich Nanorods): The Substantia Nigra Is a Very Early Target in Young Urbanites and the Gastrointestinal Tract a Key Brainstem Portal," *Environmental Research* 191 (December 2020), 110139, https://doi.org/10.1016/j.envres.2020.

[19]Katherine Kortsmit et al., "Abortion Surveillance—United States, 2018," *MMWR Surveillance Summaries* 69, no. 7 (2020): 1-29, http://dx.doi.org/10.15585/mmwr.ss6907a1.

[20]Benjamin Bowe et al., "Burden of Cause-Specific Mortality Associated with PM2.5 Air Pollution in the United States," *JAMA Network Open* 2, no. 11 (2019): e1915834, https://pubmed.ncbi.nlm.nih.gov/31747037/.

[21]Richard Fuller et al., "Pollution and Health: A Progress Update," Planetary Health, *Lancet* 6, no. 6 (2022): E535-E547, https://doi.org/10.1016/S2542-5196(22)00090-0.

[22]"First Frost: Frost-Free Season Getting Longer," *Climate Central*, October 11, 2017, https://medialibrary.climatecentral.org/resources/frost-free-season-2017.

[23]"Local: Average Date of Last Freeze," *Climate Central*, February 26, 2020, www.climatecentral.org/graphic/2020-spring-package?graphicSet=Market&location=Philadelphia&lang=en.

[24]Howard S. Ginsberg et al., "Environmental Factors Affecting Survival of Immature *Ixodes scapularis* and Implications for Geographical Distribution of Lyme Disease: The Climate/Behavior Hypothesis," *PLoS One* 12, no. 1 (2017): e0168723, https://doi.org/10.1371/journal.pone.0168723.

[25]"Nationally Notifiable Infectious Diseases and Conditions, United States: Annual Tables; Table 2J, 2019," Centers for Disease Control and Prevention, accessed November 23, 2021, https://wonder.cdc.gov/nndss/static/2019/annual/2019-table2j.html.

[26]Igor Dumic and Edson Severnini, "'Ticking Bomb': The Impact of Climate Change on the Incidence of Lyme Disease," *Canadian Journal of Infectious Diseases and Medical Microbiology*, October 24, 2018, https://doi.org/10.1155/2018/5719081.

[27]"Disease Danger Days: Days with Transmission Risk by Mosquitoes," *Climate Central*, August 8, 2018, http://ccimgs-2018.s3.amazonaws.com/2018Mosquitoes/2018Mosquitoes_detroit_en_title_lg.jpg.

[28]Alexandra Phelan and Lawrence O. Gostin, "The Human Rights Dimensions of Zika," United Nations Academic Impact, www.un.org/en/academic-impact/human-rights-dimensions-zika.

[29]Because there are instances in which abortions must be performed in order to save the life of the mother after a fetus has already died (i.e., miscarriage, fetal demise) or where fetal viability is impossible (i.e., ectopic pregnancy), a total abolition of legal abortion would be both impractical and highly cruel to women and families experiencing dangerous and/or unviable pregnancies.

[30]Jason Horowitz, "In Shift for Church, Pope Francis Voices Support for Same-Sex Civil Unions," *New York Times*, October 21, 2020, updated June 27, 2021, www.nytimes.com/2020/10/21/world/europe/pope-francis-same-sex-civil-unions.html.

[31]Adam Liptak, "A Vast Racial Gap in Death Penalty Cases, New Study Finds," *New York Times*, August 3, 2020, www.nytimes.com/2020/08/03/us/racial-gap-death-penalty.html.

6. A Story Can Change the World

[1]Anthony Leiserowitz et al., "Climate Change in the American Mind, 2021," Yale University and George Mason University for the Yale Program on Climate Change Communication, 2021, 3, www.climatechangecommunication.org /wp-content/uploads/2021/06/climate-change-american-mind-march-2021.pdf.

[2]Edward Maibach et al., "Is There a Climate 'Spiral of Silence' in America?: March 2016," Yale University and George Mason University for the Yale Program on Climate Change Communication, 2016, 6-7, https://climatecommunication.yale .edu/wp-content/uploads/2016/09/Climate-Spiral-Silence-March-2016.pdf.

[3]Maibach et al., "Is There a Climate 'Spiral of Silence' in America?," 1.

[4]Leiserowitz et al., "Climate Change in the American Mind, 2021," 17.

[5]Matthew H. Goldberg et al., "Discussing Global Warming Leads to Greater Acceptance of Climate Science," *Proceedings of the National Academy of Sciences of the United States of America* 116, no. 30 (2019): 14804-14805, https://doi .org/10.1073/pnas.1906589116.

[6]Danielle F. Lawson et al., "Children Can Foster Climate Change Concern Among Their Parents," *Nature Climate Change* 9 (2019): 458-62, https://doi.org/10.1038 /s41558-019-0463-3.

[7]Katharine Hayhoe, *Saving Us: A Climate Scientist's Case for Hope and Healing in a Divided World* (New York: One Signal, 2021), 225.

[8]"Levelized Cost of Energy, Levelized Cost of Storage, and Levelized Cost of Hydrogen," *Lazard*, October 28, 2021, www.lazard.com/perspective /levelized-cost-of-energy-levelized-cost-of-storage-and-levelized-cost-of -hydrogen/.

[9]George Marshall, workshop with the author, Paris, France, December 5, 2015. Referenced content is paraphrased.

7. God's Pleasure, Our Joy

[1]Portions of this section were originally published on CNN.com; Kyle Meyaard-Schaap, "The Key Word Missing From the Climate Movement," CNN, April 30, 2021, www.cnn.com/2021/04/30/opinions/climate-movement-joy-meyaard -schaap/index.html; used with permission.

[2]Elizabeth Kolbert, *The Sixth Extinction: An Unnatural History* (New York: Henry Holt, 2014).

[3]Per Espen Stoknes, "How to Transform Apocalypse Fatigue into Action on Global Warming," TEDGlobal Conference, New York, September 2017, www.ted.com /talks/per_espen_stoknes_how_to_transform_apocalypse_fatigue_into _action_on_global_warming.

[4]Aldo Leopold, *A Sand County Almanac* (Oxford: Oxford University Press, 1949), 197.

[5]Richard Foster, *Celebration of Discipline: The Path to Spiritual Growth* (San Francisco: HarperCollins, 1978), 7.

[6]James K. A. Smith, *You Are What You Love: The Spiritual Power of Habit* (Grand Rapids, MI: Brazos, 2016).

[7]James K. A. Smith, *Desiring the Kingdom: Worship, Worldview, and Cultural Formation* (Grand Rapids, MI: Baker Academic, 2009), 19-22.

[8]Smith, *Desiring the Kingdom*, 23.

[9]Smith, *Desiring the Kingdom*, 23.

[10]Foster, *Celebration of Discipline*, 2.

[11]*Chariots of Fire*, directed by Hugh Hudson (London: Allied Stars Ltd., 1981).

[12]Dr. Steve Bouma-Prediger, religion professor at Hope College in Holland, Michigan, published a terrific exploration of ecological virtue ethics in 2019: Steven Bouma-Prediger, *Earthkeeping and Character: Exploring a Christian Ecological Virtue Ethic* (Grand Rapids, MI: Baker Academic, 2019).

8. Loving Our Neighbors in Public

[1]Richard P. Allan et al., "IPCC, 2021: Summary for Policymakers," in *Climate Change 2021: The Physical Science Basis: Summary for Policymakers*, Contribution of Working Group I to the Sixth Assessment Report of the Intergovernmental Panel on Climate Change (New York: Cambridge University Press, 2021), 15, www.ipcc.ch/report/ar6/wg1/downloads/report/IPCC_AR6_WGI_SPM.pdf.

[2]V. Masson-Delmotte et al., "IPCC, 2018: Summary for Policymakers," in *Special Report: Global Warming of 1.5°C* (New York: Cambridge University Press, 2018), section C.2, 15, www.ipcc.ch/sr15/chapter/spm/.

[3]Masson-Delmotte et al., "IPCC, 2018," 15.

[4]Paul Griffin, "The Carbon Majors Database: CDP Carbon Majors Report 2017," CDP & Climate Accountability Institute, July 2017, www.cdp.net/en/reports/downloads/2327.

[5]John Stuart Mill, *Inaugural Address Delivered to the University of St. Andrew*, February 1, 1867 (London: Longmans, Green, Reader, and Dyer, 1867), 36.

[6]Eitan Hersh, *Politics Is for Power: How to Move Beyond Political Hobbyism, Take Action, and Make Real Change* (New York: Scribner, 2020).

[7]Aristotle, *Politics*, book 1, section 1253a.

[8]Cornel West, Twitter post, February 14, 2017, https://twitter.com/CornelWest/status/831718432995319808.

[9]"The Overton Window," Mackinac Center for Public Policy, accessed October 21, 2021, www.mackinac.org/OvertonWindow.

[10]Seth Wynes, Matthew Motta, and Simon D. Donner, "Understanding the Climate Responsibility Associated with Elections," *One Earth* 4, no. 3 (2021): 363-71, https://doi.org/10.1016/j.oneear.2021.02.008.

9. Christian Citizenship in a Warming World

[1]David Brody and Jenna Browder, "On Air Force One with President Trump: 'Nobody's Done More for Christians or Evangelicals,'" *CBN News*, November 1, 2018, www1.cbn.com/cbnnews/politics/2018/november/on-air-force-one-with-president-trump-nobodys-done-more-for-christians-or-evangelicals.

[2]Elizabeth Dias, "'Christianity Will Have Power,'" *New York Times*, August 9, 2020, www.nytimes.com/2020/08/09/us/evangelicals-trump-christianity.html.

[3]"White Evangelicals See Trump as Fighting for Their Beliefs, Though Many Have Mixed Feelings About His Personal Conduct," Pew Research Center, March 12, 2020, www.pewresearch.org/religion/2020/03/12/white-evangelicals-see-trump-as-fighting-for-their-beliefs-though-many-have-mixed-feelings-about-his-personal-conduct/.

[4]"U.S. Adults See Evangelicals Through a Political Lens," Culture & Media, Barna, November 21, 2019, www.barna.com/research/evangelicals-political-lens/.

[5]Nicholas Confessore and Karen Yourish, "$2 Billion Worth of Free Media for Donald Trump," Upshot, *New York Times*, March 15, 2016, www.nytimes.com/2016/03/16/upshot/measuring-donald-trumps-mammoth-advantage-in-free-media.html.

[6]Frederick Buechner, *Wishful Thinking: A Seeker's ABC* (San Francisco: HarperOne, 1993), 118-19.

[7]Parker J. Palmer, *Let Your Life Speak: Listening for the Voice of Vocation* (San Francisco: Jossey-Bass, 2000), 4.

[8]Palmer, *Let Your Life Speak*, 10.

[9]Debra Rienstra, *So Much More: An Invitation to Christian Spirituality* (San Francisco: Jossey-Bass, 2005), 216.

[10]Ken Untener, excerpt from a homily written for Cardinal Dearden on the occasion of the Mass for Deceased Priests, October 25, 1979, www.usccb.org/prayer-and-worship/prayers-and-devotions/prayers/prophets-of-a-future-not-our-own.